CELL AND MOLECULAR BIOLOGY IN ACTION SERIES

Human Molecular Genetics

Peter Sudbery

Longman

Addison Wesley Longman Limited
Edinburgh Gate, Harlow
Essex CM20 2JE
England
and Associated Companies throughout the World

© Addison Wesley Longman Limited 1998

First published 1998

ISBN 0 582 32266 9

British Library Cataloguing-in-Publication Data
A catalogue record for this book is
available from the British Library.

Set by 30 in Concorde BE
Produced by Addison Wesley Longman Singapore (Pte) Ltd.,
Printed in Singapore

Contents

Further information for use with this book can be accessed via a link in the catalogue entry for *Human Molecular Genetics* on the publisher's Web site at **http://www.awl-he.com/biology**. The aim of this facility is to provide updates and news of major advances in human genetics since publication. It also includes study guides for chapters, and cites additional Web site addresses which provide information of interest to human genetics. We would encourage readers to use the Web site regularly.

Acknowledgements

We are grateful for permission to reproduce copyright material in the following illustrations:

Plates 1–3 and Figure 2.8. Maltby, E., Langhill Centre for Human Genetics, Sheffield, UK; **Table 1.1** Weatherall, D.J. (1990) The New Genetics and Clinical Practice, 3rd edition, copyright of Oxford University Press; **Figure 2.9** Caskey, C.A. and Rossiter, B.J.F. (1992) *Journal of Pharmacy and Pharmocology* **44** (suppl. 1) 198–203, copyright of Charles Fry Publishers; **Figure 3.1** Schuler, G.D. *et al.* (1996) A gene map of the human genome, *Science* **274** 540–41, copyright of American Association for the Advancement of Science; **Figure 3.12** Chumakov *et al.* (1992) Continuum of overlapping clones spanning the entire chromosome 21q, *Nature* **359** 380–87, copyright of MacMillan Magazines Ltd; **Figure 3.17** from http://www-genome.wi.mit.edu/ with permission from Dr T.H. Hudson; **Figure 4.3** Collins, F. S. (1992) Cystic Fibrosis: Molecular biology and therapeutic implications, *Science* **256** 774–9, copyright of American Association for the Advancement of Science; **Figure 4.4** Nawrotski *et al.* (1996) The genetic basis of neuromuscular disorders, *Trends in Genetics* **12** 294–8, copyright of Elsevier Science Ltd; **Figure 4.5** The Huntington's disease collaborative research group (1993) A novel gene containing a trinucleotide repeat that is expanded and unstable on Huntington's disease chromosomes, *Cell* **72** 971–83, copyright of Cell Press; **Box 4.2** Marx, J.L. *et al.* (1989) The cystic fibrosis gene is found, *Science* **245** 923–5, copyright of Dr L-C. Tsui; **Figure 5.4** Cordell, H.J. and Todd, J.A. (1995) Multifactorial inheritance of type I diabetes, *Trends in Genetics* **11** 499–503 copyright of Elsevier Science Ltd and P.E. Applied Biosystems; **Figures 5.6 and 5.7, and Table 5.2** Bennet, S.T. and Todd, J.A. (1996) Human Diabetes and the Insulin Gene: Principle of mapping polygenes, *Annual Review of Genetics* **30** 343–70, copyright of Annual Reviews Inc; **Figures 7.1, 7.2, 7.3, 7.4, 7.5, 7.6, 7.9, 7.11** North Trent Molecular Genetics Laboratory, Sheffield, UK; **Figure 9.1**

Efstratiadis, A. *et al.* (1980) The structure and evolution of the human β globin family, *Cell* **21** 653–68, copyright of Cell Press; **Figures 9.3, 9.4 and 9.9 (top and middle)** Cavalli Sforza, L.L. *et al.* (1996) The history and geography of human genes, abridged paperback edition, copyright of Princeton University Press; **Figure 9.8** Armour, J.A.L. *et al.* (1996) Minisatellite diversity supports a recent African origin for modern humans, *Nature Genetics* **13** 154–1601, copyright of Nature America Inc; **Figure 9.9 (bottom)** Devoto *et al.* (1990) Gradient of distribution in Europe of the world CF mutation and its associated haplotypes, *Human Genetics* **85** 436–41, copyright of Springer Verlag; **Figure 10.3** from http://www2.perkin-elmer.com:80/fo/773201/773201.html copyright of PE Applied Biosytems.

We would especially like to thank the following for their valuable reviews on draft forms of the manuscript:

John Armour (University of Nottingham)
Kerry Bloom (University of North Carolina)
Keith Brown (University of Bristol)
Dave Curtis (Royal College of London)
Ed Wood (University of Leeds)

Preface

This book grew out of a third-year module I teach in the genetics degree at Sheffield University. During the preparation of the course I was impressed by the momentous importance of recent progress in human genetics, both at a fundamental level and for its likely impact on medical practice. My aim in this book has been to make this field accessible to as wide an audience as possible by providing a route map through a conceptually demanding and fast-moving field. The intended readership is second- and third-year undergraduates studying genetics or a related subject. However I hope a much wider audience will find at least part of the book relevant to their work. This may include professional scientists in other research fields, medical students and those training in the other healthcare professions, school teachers in sixth-forms, the secretariat of regulatory and bioethics committees, lawyers active in biotechnology and managers in the biotechnological industries.

The first two chapters generally set the scene for an account of recent work in the following eight chapters, concluding with a discussion of the impact of human genetics on society. More advanced readers may wish to commence at Chapter 3. In order to keep the book reasonably concise and inexpensive I have assumed that the reader will be familiar with the general concepts of genetics and molecular biology such as would normally be taught in the first year of a university course. These topics already receive excellent coverage in many general textbooks and there seems little point in repeating what is already widely available. Some of the material in Chapter 2 in particular is dealt with in more depth in these textbooks. Consequently, I hope that this book provides a more focused and more up-to-date account of the systematic analysis of the human genome than is found in a general textbook. To help the reader whose background knowledge may not be sufficient I have provided a glossary of concepts, specialised terminology, abbreviations and acronyms. For the most part, words highlighted in bold in the main text are defined in the glossary. However, some additional words and concepts are also defined. For those readers who wish to investigate particular topics in more detail, each chapter finishes with a guide to further reading that includes many excellent reviews with which the field is well served. I also recommend the excellent textbook by Professors Strachan and Read (*Human Molecular Genetics*, Strachan, T. and Read, A.P. (1996). Bios Scientific Publishers, Oxford, UK), which covers many topics in greater depth than would be sensible within the brief of this book. I have tried to make each chapter as self-contained as possible, for I believe that only the most committed start at page one and read serially through to the last page. A more common pattern is that the reader wishes to be informed about particular topics, for example gene therapy or genome mapping. This has occasionally necessitated a small amount of repetition.

An introductory text to such a large and complex field must restrict itself to illustrating general principles with selected examples. Inevitably work on many important genes must be omitted, and I can only apologise to those who feel their favourite gene has not received the attention it deserves. Equally, a certain amount of simplification is essential to make it accessible to an undergraduate audience. This is particularly true of those areas that are mathematically based, which experience shows can be a cause of difficulty to biology students. Again, I can only hope that I have succeeded in spreading enlightenment without committing sins of oversimplification.

Anyone who researches the literature will quickly find that figures quoted for the incidence of genetic disease vary widely. I have mainly relied on figures in *The New Genetics and Clinical Practice* by Professor David Weatherall and the House of Commons Report on Human Genetics (full details of both references are given at the end of Chapter 1). On occasion I have used figures from published work that report the data on which frequency estimates are based, e.g. population frequencies of mutations in the *BRCA*1 and *BRCA*2 genes that predispose to breast cancer.

I have tried to keep the account and the references at the end of each chapter as up-to-date as possible. I will also be maintaining a Web page to inform the reader of developments after the book has been published. I welcome feedback by email concerning the book and of recent developments that would be suitable for the page.

I could not have imagined before I commenced work the extent to which I would be beholden upon the goodwill and help of others. I hope the following will adequately convey my heartfelt thanks to the many people who helped me with this book. Firstly, to my wife Carol for her moral support, her tolerance of my neglect of family duties while writing and the many hours she spent reading drafts. To my laboratory group who remained remarkably tolerant of my lack of attention. To Katherine Duffy for her diligent proof-reading. To Alex Seabrook and Tina Cadle at Addison Wesley Longman who at all times were a model of professionalism, tolerant of my failings and generally provided superb support throughout. To Anne Dalton, Steve Evans and Diana Curtis at the North Trent Molecular Genetic Laboratory and Edna Maltby at the Langhill Centre for Human Genetics who generously provided samples of their work. Lastly, I owe a particular debt to John Armour (University of Nottingham), Kerry Bloom (University of North Carolina), Keith Brown (University of Bristol), Dave Curtis (Royal London Hospital) and Ed Wood (University of Leeds) whose detailed, critical reading of the manuscript was an essential part of any merit the book may be deemed to have. Any remaining errors, inaccuracies or other failings are, of course, entirely my own responsibility.

Peter Sudbery

Abbreviations

AAV	adeno-associated virus
ABC	ATP-binding cassette
ACGT	Advisory Committee for Gene Testing
ADA	adenine deaminase deficiency
AGE	agarose gel electrophoresis
ARMS	amplification refractory mutation system
ASO	allele specific oligonucleotide
AT	ataxia telengiectasia
BAC	bacterial artificial chromosome
BMD	Becker muscular dystrophy
CBAVD	congenital bilateral absence of the vas deferens
CDK	cyclin dependent kinase
CEPH	Centre d'Étude Polymorphism Humain
CF	cystic fibrosis
CFTR	cystic fibrosis transmembrane conductance regulator
CMC	chemical mismatch cleavage
CMD	congenital muscular dystrophy
CPEO	chronic progressive external opthalomoplegia
CRE	cyclic AMP response element
DASC	dystrophin associated sarcoglycan complex
DGGE	denaturing gradient gel electrophoresis
DMAHP	dystrophia myotonica associated homeobox protein
DMD	Duchenne muscular dystrophy
DMPK	dystrophia myotonica protein kinase
DOE	Department of the Environment (USA)
ELSI	ethical, legal and social issues
EMBL	European molecular biology laboratory
EMC	enzyme mismatch cleavage
EPC	European patent convention
EST	expressed sequence tag
FAP	familial adenomatous coli
FISH	flourescence in situ hybridisation
GAP	GTP-activating protein
GTAC	gene therapy advisory committee
HD	Huntington's disease

HGAC	Human Genetics Advisory Committee	NIH	National Institute of Health (USA)
HLA	human leucocyte antigens	NOD	non-obese diabetic (mouse)
HNPCC	hereditary non-polyposis colorectal cancer	ORF	open reading frame
HPFH	hereditary persistence of foetal haemoglobin	PAC	P1-derived artificial chromosome
HRE	heat response element	PC	principle component
HUGO	Human Genome Organisation	PCR	polymerase chain reaction
IBD	identical by descent	PFGE	pulse field gel electrophoresis
IBS	identical by state	PIC	polymorphism information content
IDDM	insulin-independent diabetes mellitus	rDNA	DNA coding for ribosomal RNA molecules
LCR	locus control region	RFLP	restriction fragment length polymorphism
LHON	Leber's hereditary optic neuropathy	RH	radiation hybrid
LINE	long interspersed element	SCID	severe combined immune deficiency syndrome
LOD	log ratio of odds	SINE	short interspersed element
LOH	loss of heterozygosity	snRNP	small nuclear ribonucleoprotein
LTR	long terminal repeat	SRE	serum response element
MD	myotonic dystrophy	SSCP	single stranded conformational polymorphism
MDR	multiple drug resistance		
MELAS	mitochondrial encephalomyopathy, lactic acidosis and stroke-like episodes	STC	sequence tagged connector
		STR	short tandem repeat
MHC	major histocompatibility complex	STS	sequence-tagged factor
MLS	maximum log score	TBF	TATA-box binding factor
MODY	maturity onset diabetes of the young	TDT	transmission disequilibrium test
mtDNA	mitochondrial DNA	TNF	tumour necrosis factor
MVR-PCR	multivariant repeat-polymerase chain reaction	TRE	trinucleotide repeat expansion
NARP	Neurogenic muscle weakness, ataxia and retinitis pigmentosa	UPGMA	unweighted pair group method with arithmetic mean
NBF	nucleotide binding fold	URL	uniform resource indicator
NCHGR	National Council for Human Genetic Research	UTR	untranslated region
NF	neurofibromatosis	VNTR	variable number tandem repeat
NIDDM	non-insulin independent diabetes mellitus	YAC	yeast artificial chromosome

Human genetic disease

Key topics

- Frequency and types of genetic disease
- Single-gene disorders
  - Complexity in single-gene disorders
  - Autosomal recessive
  - Autosomal dominant
  - Sex-linked
- Multifactorial or complex disorders
  - Evidence for genetic factors in common diseases
  - Genetic influences on personality disorders and phenotypic traits
- Chromosomal imbalances
- Mitochondrial disorders

1.1 Introduction

About 5% of liveborn babies suffer from a significant medical disorder, which may be life-threatening or, at the very least, require hospital treatment. A disorder present at birth is said to be **congenital**. Most congenital disorders will have a genetic component in their **aetiology**. This may take the form of a single-gene or monogenic defect, a mitochondrial disorder, a chromosomal imbalance or a multifactorial condition, which is partly genetic and partly environmental. Table 1.1 shows estimates of the genetic burden affecting the health of the newborn imposed by these various genetic factors.

A survey of admissions to a paediatric hospital in Montreal showed that one-third of all admissions were for diseases with a genetic component. Moreover, 70% of patients admitted more than once had a disorder with a genetic component. Such disorders are responsible for an immense amount of suffering, reduced quality of life, shortened life expectancy and distress to

Table 1.1 The genetic load in the newborn. It is difficult to give precise figures because of variation between different populations, so the figures are in the form of ranges. The genetic load of congenital defects is estimated as half the overall incidence of such disorders. No overall figures are available for mitochondrial disorders.

Category	Frequency (per 1000 live births)
Single gene	
Autosomal dominant	1.8–9.5
Autosomal recessive	2.2–2.5
X-linked	0.5–2.0
Chromosomal	6.8
Congenital malformations	19–22

family members and carers. Moreover, it is becoming increasingly clear that many of the common diseases of later life, perhaps up to two-thirds, have a genetic component in their aetiology. Clearly our genetic constitution plays a major part in determining our lifetime health. This chapter considers the ways in which this comes about and introduces some of the major genetic disorders that are considered in more detail in later chapters.

1.2 Single-gene defects

Single-gene or **monogenic** disorders, as their name implies, are traceable to a defect in a single gene. They follow simple patterns of inheritance, predictable from the Mendelian laws of genetics. These patterns are classified according to whether the affected gene is located on the X chromosome or one of the 22 autosomes and whether the trait is recessive or dominant. That is, they are said to be **X-linked**, **autosomal recessive** or **autosomal dominant**, respectively. The Online Mendelian Inheritance in Man database (see Further reading at the end of the chapter) currently lists over 5000 disorders that have been definitely traced to single-gene defects. This represents a significant fraction of the total number of human genes, estimated to be between 65 000 and 80 000 (see Chapter 2). Single-gene disorders that are both severe and relatively common are known as **major disorders**.

The frequency of mutant alleles responsible for a monogenic disorder in a population is specified by the **Hardy–Weinberg distribution**, which states that:

$$p^2 + 2pq + q^2 = 1$$

where p is the allele frequency of the more common allele and q the frequency of the less common allele. This equation is only true if certain conditions are met, such as mating is random and there is no migration into or from the population. This is discussed in more detail in Chapter 9. From this equation the carrier (heterozygote) frequency for recessive disorders can be calculated ($2pq$) from the observed frequency with which the disease occurs in a population. This can give surprising results. For example, the frequency of cystic fibrosis in the UK is 1 in 2000, from which the carrier frequency can be calculated as 1 in 22.

1.2.1 Autosomal recessive disorders

Autosomal recessive diseases require the inheritance of two defective alleles. This means that there are no functioning copies of the gene and implies that the disorder results from a loss of function. An example of an autosomal recessive pattern of inheritance is shown in Figure 1.1 and the frequencies of some major autosomal recessive diseases are shown in Table 1.2.

Typically more than one sibling of unaffected parents may be affected; children of cousin marriages are also at risk (Figure 1.1). Both parents must be carriers or **heterozygous** at the locus concerned (Figure 1.2); as a result, for each child there is a 25% chance that it will be affected, a 50% chance that it will be a carrier and a 25% chance that it will be completely normal. If the normal allele is dominant, the ratio of unaffected to affected children will be 3:1. To put it another way, for each child born there is a 25% chance that it will be affected.

As shown in Figure 1.1, the children of a cousin marriage may be homozygous for a deleterious allele that is heterozygous in a common grandparent. In fact, the chance that this will happen is 1 in 64 for each such allele. Since everyone is likely to be heterozygous for about five different deleterious alleles, cousin marriages are more likely to result in children affected by an autosomal recessive disorder than marriages between

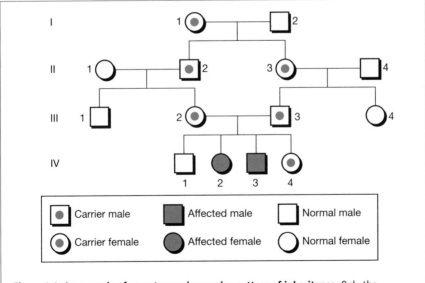

Figure 1.1 An example of an autosomal recessive pattern of inheritance. Only the **homozygous** recessive individuals (IV-2 and IV-3) are affected. Note how the heterozygous state in I-1 becomes homozygous in IV-2 and IV-3 as a result of the **consanguineous** marriage of III-2 and III-3.

Table 1.2 Some major monogenic recessive disorders.

Disease	Frequency	Symptoms
Cystic fibrosis	1 in 2000 (N. Europeans)	Recurrent lung infections, pancreatic exocrine deficiency, male sterility
α_1- Antitrypsin deficiency	1 in 5000 to 1 in 10 000 (N. Europeans)	Liver failure, emphysema
Phenylketonuria	1 in 2000 to 1 in 5000 (Europeans)	Mental retardation
Tay-Sachs disease	1 in 3000 (Ashkenazi Jews)	Neurological degeneration, blindness and paralysis
Sickle cell anaemia	1–2 in 100 (Africa where malaria is endemic)	Anaemia
Haemochromatosis	1 in 500	Excessive iron accumulation in adults, resulting in diabetes, liver cirrhosis and heart failure
Thalassaemias	1–2 in 100 (Mediterranean and Asia where malaria is endemic)	Anaemia

unrelated parents. Indeed, many rare monogenic disorders are only observed in the children of cousin marriages.

Some diseases due to single-gene defects are present in certain populations at a much higher frequency than would be expected by mutation alone. Important examples of this are cystic fibrosis in European populations, sickle cell anaemia and thalassaemias in Asian and African populations, and Tay–Sachs disease in **Ashkenazi Jews**. One reason for an elevated frequency

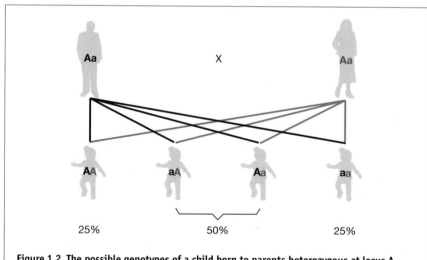

Figure 1.2 The possible genotypes of a child born to parents heterozygous at locus A.

is that the heterozygote may enjoy some advantage, resulting in the allele frequency being elevated by selection. In the case of sickle cell anaemia or thalassaemias, it has been demonstrated that heterozygotes are more resistant to malaria, which is endemic in those countries where such disorders are prevalent. This variation means that the ethnic origin of a patient may be relevant in arriving at a correct diagnosis. Other possible reasons for elevated disease frequencies are discussed in Chapter 9.

Cystic fibrosis

Cystic fibrosis (CF) is a common single-gene disorder among people of European extraction. It affects approximately 1 in 2000 babies, and about 1 in 22 are carriers. The affected gene encodes a protein known as the **cystic fibrosis transmembrane conductance regulator** (CFTR; see Chapter 4). The CFTR protein is responsible for the export of chloride ions across the plasma membrane of epithelial cells that line the lung airways. CF alleles impair or obliterate this function and, since water molecules follow the ions due to osmosis, insufficient water is secreted on to the cell surface. This results in a thick and sticky mucus that causes congestion in the lung airways. In a healthy person, the cilia of the epithelial cells continually move the mucus to the top of the airways and into the digestive system. In this way foreign bodies including bacteria are removed from the lungs. In a CF patient, the mucus is too thick to be moved by the cilia. Consequently this cleansing mechanism fails, resulting in repeated bacterial infections. The cumulative effect of these causes long-term damage to the lungs, and even with the best treatment the maximum life expectancy is 30 years. CF also affects other bodily systems. The pancreatic duct may become blocked, resulting in digestive difficulties. This is known as pancreatic exocrine deficiency. In some male patients, the vas deferens does not form properly, resulting in sterility. This is known as **congenital bilateral absence of the vas deferens** (CBAVD); sometimes this can occur without any other obvious symptoms of CF. Finally, CF results in an excess of chloride ion secretion in sweat glands, resulting in abnormally salty sweat. This can be simply recognised by measuring sweat electrolyte levels.

The haemoglobinopathies

Sickle cell anaemia, α and β **thalassaemias** and **glucose 6-phosphate dehydrogenase deficiency** (in some environments) affect the formation of haemoglobin, causing a class of disease known as the haemoglobinopathies. Haemoglobin is a tetramer of two β-globin molecules and two α-globin molecules, each complexed with a molecule of haem. Sickle cell anaemia results from a single-base mutation affecting the gene encoding β-globin. Study of this disease played an important part in the development of modern molecular genetics, because in 1956 Ingram demonstrated that the β-globin of sickle cell patients differed from the normal protein by one amino acid, providing experimental evidence for the first time that the sequence of amino acids in proteins was determined by genes. It is also important because it was the first, and still the clearest, example in humans of heterozygous advantage. Carriers of the sickle cell gene are more resistant to malaria. This results in selection for the mutation and consequently an increased occurrence of the recessive homozygotes who suffer from the disease. Such a situation is known as a balanced polymorphism. The α and β thalassaemias result in a decrease or absence of α-globin and β-globin respectively. Like sickle cell anaemia, heterozygotes are more resistant to malaria so a balanced polymorphism results in thalassaemias being more common where malaria is, or was, endemic.

Tay–Sachs disease

Tay–Sachs disease is a progressive neurological degeneration that starts in the first year of life, characterised by developmental and mental retardation, progressive muscle weakness and paralysis, and blindness. Death usually results by the age of 5 years. It is caused by a mutation in the *HEXA* gene, which encodes the α-subunit of hexosaminidase A, a lysosomal enzyme required for the breakdown of a complex **glycolipid** called **ganglioside** GM_2 to a simpler molecule called ganglioside GM_3. The build-up of ganglioside GM_2 impairs neurone function and results in the disease. It is unusually common in the Ashkenazi Jew population, where it affects 1 in 3600 of the population.

Phenylketonuria

Phenylketonuria is characterised by mental subnormality. It results from a lack of phenylalanine hydroxylase, which converts phenylalanine to tyrosine, the first step in the phenylalanine degradation pathway. Phenylalanine consequently accumulates in the bloodstream. The damage to the developing nervous system is actually caused by phenylpyruvic acid, to which phenylalanine spontaneously converts. Most babies in the UK and the USA are now screened at birth for this deficiency because it can be almost completely controlled by a low-phenylalanine diet. It is thus an interesting example of a disorder that may be either completely genetically determined or completely environmentally determined.

α_1-Antitrypsin deficiency

Antitrypsin is an inhibitor of elastase. Deficiency results in unregulated breakdown of connective tissue that particularly affects the elasticity of the lungs, resulting in emphysema. It is another disorder that is particularly common among people of European descent. The protein can be readily made by recombinant methods, for example transgenic sheep have been produced that secrete large amounts of the protein in their milk. This provides the prospect of a cheap and effective treatment of the disease.

Haemochromatosis

This is probably the most common recessive disorder in the UK. It is caused by excessive iron accumulation and can be treated by blood letting. The symptoms are very variable and it is often not correctly diagnosed. It commonly affects women after the menopause, because the blood loss during menstruation is protective.

Autosomal dominant disease results from the inheritance of only one mutant allele. This usually results in a protein that has gained a novel function or is expressing its normal function in an unregulated fashion. If affected individuals are able to reproduce, the disorder is likely to be manifested in every generation of a pedigree in which it is segregating. Thus 50% of the children of an affected person are likely to suffer from the disease (Figure 1.3).

If the disease is sufficiently serious to prevent reproduction, it follows that most cases will arise from *de novo* mutation during **gametogenesis** and the occurrence of the disease will be apparently sporadic. Some autosomal dominant disorders are only manifested later in life, after an individual is likely to have finished reproducing. Such mutations will escape the effect of selection. Examples of this are Huntington's disease, familial breast cancer (see Chapter 4), and hereditary Alzheimer's disease (see Chapter 5). Some examples of major autosomal dominant disorders are shown in Table 1.3.

Huntington's disease

Huntington's disease (HD) is a progressive neurological degeneration that affects patients in middle and later life from their fifth decade onwards. Typical symptoms include dementia, severe depression and a characteristic involuntary, dance-like movement known as chorea. It is a particularly distressing condition, because by the time the symptoms are evident an affected person is likely to have already had children, which each have a 50% chance of suffering from the disease. The HD gene proved difficult to clone and this was only achieved in 1993. The mutation that causes the dis-

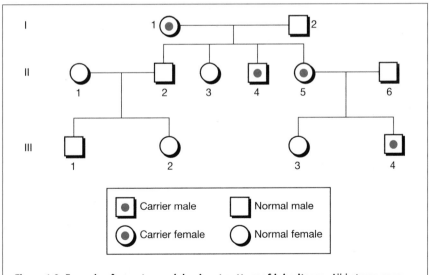

Figure 1.3 Example of an autosomal dominant pattern of inheritance. All heterozygous individuals are affected (I-1, II-4, II-5 and III-4). Since 50% of children will be affected, the disease is likely to be manifested in each generation.

Table 1.3 Example of some major autosomal dominant disorders.

Disorder	Frequency	Symptoms
Familial hypercholesterolaemia	1 in 500	Premature heart disease
Familial breast cancer		
BRCA1	1 in 800 (USA)	High lifetime risk of breast and
BRCA2	1 in 100 (Ashkenazi Jews[1])	ovarian cancer. Earlier onset than sporadic cases
Familial Alzheimer's disease	10% lifetime risk at age 80[2]	Dementia. Earlier onset than sporadic cases
Hereditary non-polyposis colorectal cancer	1 in 400[3]	Colon cancer not associated with polyps
Familial adenomatous polyposis	1 in 8000	Bowel polyps that may become malignant
Neurofibromatosis type 1	1 in 4000	Tumours of peripheral nerves. *Café-au-lait* pigment spots on skin
Huntington's disease	1–2 in 10 000	Involuntary choreiform movements. Dementia. Late onset
Myotonic dystrophy	1 in 8500	Myotonia, heart defects and cataracts
Familial retinoblastoma	1 in 14 000	Multiple unilateral and bilateral tumours of the retina

[1] *BRCA1* and *BRCA2* each have a frequency of 1 in 100 in Ashkenazi Jews.

[2] Alzheimer's disease is normally a complex disease with genetic and environmental components. A subset of cases show early onset and simple autosomal dominant inheritance. What proportion of total cases are represented in this category is not known.

[3] Wide range of estimates from 1 in 200 to 1 in 10 000. (see section 4.8.4).

ease is known as a trinucleotide repeat expansion (see Chapter 4). In normal individuals a sequence is found in which the triplet CAG is repeated about 15 times (the actual number can vary a small amount around this value). In HD patients the number of repeats increases to 36 or more. This type of mutation has been found in a number of other genes that have been cloned recently. The existence of this type of mutation was entirely unexpected and could only have been discovered by the cloning and characterisation of the genes involved.

Familial breast cancer
Genetic influences are not thought to be an important factor in the occurrence of most breast cancer cases (see below). However, a subset, perhaps about 5% of all cases, follow a pattern of inheritance characteristic of an autosomal dominant mutation. Two genes, *BRCA1* and *BRCA2* have been identified and cloned (see Chapter 4). They were recognised by studying cases of breast cancer that occurred before the patient was 40 years old, an unusually early age for the onset of breast cancer. Inheritance of *BRCA1* results in

80% lifetime risk of breast cancer. As well as leading to predisposition to breast cancer, both *BRCA1* and *BRCA2* (to a lesser extent) also cause a predisposition to other cancers, particularly ovarian cancer. Inheritance of the *BRCA2* allele in males carries a 1 in 100 risk of male breast cancer. Although responsible for only a small fraction of breast cancer cases, *BRCA1* is a relatively common disease allele, estimated in US populations to occur at a frequency of 1 in 800. In Ashkenazi Jewish populations the frequency for both *BRCA1* and *BRCA2* is much higher (1 in 100).

Myotonic dystrophy

Myotonic dystrophy (MD) is characterised by myotonia (delayed muscle relaxation) and degeneration of various organs including heart and eyes. It is variable in its expression, ranging from minimally affected late onset through classic adult onset to congenitally affected children of affected mothers. Although characterised as a Mendelian disorder of the autosomal dominant type, its pattern of inheritance shows some deviation from the classic pattern. In particular it may show **anticipation**, where a mother may be only slightly affected but have a severely affected child. Often the mother's symptoms are so mild that the disease is only diagnosed when her child is found to be affected. MD is another example of a trinucleotide repeat expansion. It is now known that anticipation is a result of the instability of the expansion (see Chapter 4). MD is a relatively common disorder, the allele frequency being 1 in 8500.

1.2.3 X-linked disorders

X-linked disorders affect genes located on the X chromosome. Because males only have one copy of this chromosome and therefore one copy of each of the genes located on it, they are much more likely to suffer from such disorders. Thus X-linked disorders affect male children of unaffected parents. Most X-linked disorders are recessive, so for the most part females are unaffected but act as carriers. Thus in a pedigree in which a sex-linked mutation is segregating, the mutation is inherited through the female line and the resultant disorder can apparently skip generations (Figure 1.4). Another characteristic of X-linked disorders is that uncles and nephews are often affected. One rare documented example of an X-linked dominant condition is hypophosphataemic (vitamin D-independent) rickets. In this case an affected father always passes the disorder to his daughters, but never to his sons.

Of course, it is possible for a female to be homozygous for an X-linked mutation and therefore suffer from the disease. Sometimes, a heterozygous female may show some symptoms of the disease. This may happen because of **X-chromosome inactivation** or **Lyonisation**. Named after its discoverer, Mary Lyon, this is a process that occurs early in development to randomly inactivate one or other of the X chromosomes. Cells descended from this progenitor cell abide by the decision made. Thus the body of a heterozygous female is a mosaic, consisting of patches of different clones, some of which will lack the product of the defective gene. If the function is cell autonomous, i.e. the function cannot be supplied by another cell expressing

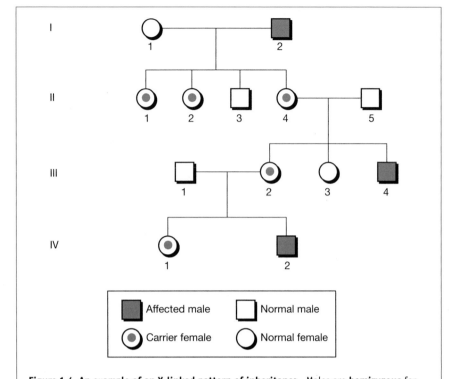

Figure 1.4 An example of an X-linked pattern of inheritance. Males are **hemizygous** for the X chromosome and have only one copy of genes on the X chromosome; males that inherit the affected gene are shown with a filled square. Note that the great-grandfather of IV-2 was affected, but not his mother or grandmother. The disease therefore skipped generations II and III with respect to IV-2 and the disorder was passed through the female line. However, IV-2's uncle was also affected, a typical pattern in X-linked disorders. Finally, note that daughters of affected males are obligate carriers.

the gene, it is possible that the lack of function will affect the overall phenotype of the carrier. This may be the reason why, for example, one-third of female carriers of fragile-X syndrome are mildly affected (see below).

The major X-linked disorders are shown in Table 1.4.

Table 1.4 Some examples of X-linked disorders.

Disorder	Frequency	Symptoms
Haemophilia A and B	1 in 10 000	Abnormally prolonged bleeding after trauma
Duchenne muscular dystrophy	1 in 3000 to 1 in 4000	Muscle wastage in teenage years
Fragile-X syndrome	1 in 1000	Mental retardation

Haemophilia
Haemophilia is an inability to form blood clots, resulting in prolonged bleeding upon injury and spontaneous internal bleeding. At a molecular

level it results from a lack of either blood clotting factor VIII (haemophilia A, 80% of cases) or blood clotting factor IX (haemophilia B or Christmas disease). It is famous for affecting the Victorian royal family.

Duchenne muscular dystrophy
Duchenne muscular dystrophy (DMD) results in progressive muscle wasting or dystrophy starting in teenage years. It soon results in confinement to a wheelchair and affected boys normally die in their twenties. There is a milder form called Becker muscular dystrophy which maps to the same gene. DMD is notable for the size of the gene affected, 2.5 Mb, equivalent to 60% of the entire genome of *Escherichia coli* (see Chapter 4). However, after splicing the mRNA is 14 kb in size; thus 99% of the gene is in the form of introns and only 1% actually encodes amino acid sequence.

Fragile-X syndrome
Fragile-X syndrome results in mental subnormality. After Down's syndrome, it is the most common cause of mental subnormality in boys and is said to be responsible for the excess of boys in institutions for the mentally handicapped. It derives its name from a cytogenetic observation: when cells are cultured in medium that is deficient in precursors of DNA metabolism the X chromosome appears to have a break in metaphase spreads. At least five X-linked loci have been identified, FRXA–FRXE. FRXA has been shown to be due to a trinucleotide repeat expansion of the type discussed above for MD and HD (see Chapter 4). Fragile-X syndrome mildly affects one-third of female carriers, so is partially dominant.

1.2.4 Complexity in single-gene disorders

There could be a temptation to regard analysis of monogenic disorders as straightforward, i.e. that the sole factor in their occurrence is the disease gene, whose inheritance follows simple and predictable patterns. In fact, this is often very far from the case. There are many factors that may modify the pattern of inheritance or symptoms of the disorder and these are discussed below. Such complexity provides a warning of the intricacies that may be encountered in the analysis of multifactorial disorders, which involve a combination of **polygenic** and environmental factors.

Genetic heterogeneity
This describes a situation where apparently clinically similar disorders are caused by mutations in different genes (non-allelic) or where mutations in the same gene result in clinically diverse conditions (allelic). Some examples of each are given in Table 1.5.

Penetrance
This refers to the frequency with which the disorder or phenotype is manifested in an individual who has inherited the disease allele. A mutation that does not inevitably cause the disorder is said to show incomplete pene-

trance. An example is split hand syndrome, a claw-like deformity of the hand caused by an autosomal dominant allele. Some individuals inherit the allele, but have normally formed hands.

Expressivity

This describes differences in the severity of a disorder in individuals who have inherited the same disease alleles. In the case of sickle cell anaemia, in which all cases result from mutation affecting the same amino acid, symptoms may be sufficiently severe to cause childhood death or be so mild that the disease remains undiagnosed until middle age. One possible cause of variation is that the effect of the sickle cell mutation is modified by other genes. In the case of sickle cell anaemia, one such modifying gene has been mapped to the X chromosome.

Mosaicism

Mosaicism arises when not all cells in the body are genetically identical. This may come about through a mutation in early development and may result in either the germline or somatic cells being affected. If the germline is mosaic, gametes may arise from progenitor cells of different genetic constitutions. For example, in one line the progenitor cell may be heterozygous for an autosomal dominant mutation, while in another the progenitor cell may be completely normal. This will clearly disturb the proportion of gametes affected, expected to be 50% if the parent is heterozygous for an autosomal dominant allele. Note that the somatic cells of females heterozygous for a sex-linked mutation will necessarily be mosaic as a result of X-chromosome inactivation.

Table 1.5 Some examples of genetic heterogeneity.

Allelic	Non-allelic
Muscular dystrophy	Profound deafness
Becker	Retinitis pigmentosa
Duchenne	Autosomal dominant
Cystic fibrosis	Autosomal recessive
Pancreatic sufficiency/deficiency	X-linked
CBAVD	Polycystic kidney disease

Phenocopy

Sometimes an environmental factor may result in a disorder with the same symptoms as an inherited disorder. For example, the infection of a mother during pregnancy with the rubella virus may result in a profoundly deaf baby, a condition that can also be caused by a number of different single-gene defects. This is known as phenocopy.

Environmental effects

Environmental factors may influence the penetrance or expressivity of a disease allele. A clear example of this is phenylketonuria, a disease that may be largely prevented by a low-phenylalanine diet.

Anticipation

There are a number of diseases whose severity apparently increases with each succeeding generation. MD and fragile-X syndrome are two well-known examples. Often the disorder may be so mild in the parent that it has not been diagnosed prior to its occurrence in a child. This phenomenon is known as anticipation.

Genomic imprinting

Genomic imprinting is said to occur when the expression of an allele depends on the parent from which it was inherited. This has been revealed in a number of different situations. An example is the effect of parental origin of a deletion of chromosomal band 15q12. Individuals heterozygous for this deletion suffer from a different disease according to the parent from which it was inherited. If the deletion was inherited from the father Prader–Willi syndrome results, characterised by mental retardation, hypotonia, gross obesity and hypogenitalism. If the deletion was inherited from the mother Angelman syndrome results, characterised by mental and growth retardation, hyperactivity and inappropriate laughter. This suggests that the function of the remaining allele depends on its parental origin. The mechanism is thought to involve DNA methylation, but the details are currently unclear.

1.3 Multifactorial or complex disorders and traits

It is a commonplace observation that members of the same family are likely to resemble each other more than they resemble the general population and are more likely to suffer from the same diseases. Since family members are likely to share the same environment as well as the same genes, it is not easy to disentangle the relative contributions of genetic and environmental factors that may determine what we are like and what diseases we are likely to suffer from. For the most part, susceptibility to common diseases and phenotypic characteristics do not follow simple patterns of inheritance, so neither is likely to be determined monogenically. Nevertheless, family, adoption and twin studies clearly show a strong genetic component. A list of common disorders in which a genetic component has been demonstrated is shown in Table 1.6. This list encompasses such a broad range of diseases that a patient's genetic constitution may be considered a factor in the majority of medical conditions.

These diseases are often referred to as **multifactorial diseases** because there are both genetic and environmental factors responsible for the onset of the disease. The ways this may occur are discussed in more detail in Chapter 5. At this point it is sufficient to note that the interactions are likely to be complex because they involve the interactions of several genes (polygenes or oligogenes) with each other and with the environment. For this reason they are commonly referred to as **complex diseases**. Because of this complexity, it is unlikely an individual's health prospects can ever be predicted in a deterministic fashion from genetic tests. However, genes that act as risk factors are

Table 1.6 Examples of common disorders for which there is evidence of a genetic component.

Congenital disorders	Common diseases of later life	Psychiatric disorders
Neural tube	Rheumatoid arthritis	Manic depression
Spina bifida	Various cancers	Alcoholism
Anencephaly	Epilepsy	Schizophrenia
Congenital heart disease	Multiple sclerosis	Tourette's syndrome
Cleft lip and palate	Insulin-dependent	Dyslexia
Mental retardation	diabetes mellitus	Alzheimer's disease
	Non-insulin dependent	
	diabetes mellitus	
	Peptic ulcer	
	Ischaemic heart disease	
	Hyperthyroidism	
	Gallstones	
	Migraine	
	Asthma and other allergies	

being identified for a wide range of common diseases (see Chapter 5). These may well serve as the basis for tests to determine an individual's predisposition or susceptibility to common disorders, allowing early diagnosis and consequently more successful treatment if the disease should occur. Moreover, defining genetic variables simplifies the analysis of environmental factors. Thus genetic tests for predisposition to common diseases may allow at-risk individuals to change their lifestyle to avoid contracting the disease.

1.3.1 Evidence for a genetic component in complex disorders and phenotypic traits

Family studies

Members of a family share a proportion of genes in common that may be simply predicted from their degree of relatedness. For example, siblings or parent–child pairs share 50% of their genes. Pairs of individuals in an extended family may be classified as first-, second- or third-degree relatives according to the proportion of genes they share (Table 1.7). If genetic factors play a role in the occurrence of the disease, then the average risk that a relative of an affected individual will also suffer from the same disease should be correlated to the degree of relatedness to the affected person. Table 1.8 shows two examples of congenital disorders where this has been shown to be true.

Adoption studies

One way to dissect the different contributions of nature and nurture is to measure the recurrence of a disease in biological relatives who have been reared apart through adoption. This may then be compared with the recurrence risk in a control group of non-adoptive families. For example, the risk

Table 1.7 Genes shared in common by different classes of relatives.

Class of relative	Proportion of genes shared	Examples
First degree	50%	Parent–child, siblings
Second degree	25%	Grandparent–grandchild, aunt or uncle–niece or nephew
Third degree	12.5%	Cousins, great-grandparent–great-grandchild

Table 1.8 Observed risks to relatives of probands suffering from congenital malformations.

Category affected	Cleft lip	Pyloric stenosis
General population	0.001	0.001
First-degree relatives	×40	×10
Second-degree relatives	×7	×5
Third-degree relatives	×3	×1.5

that the offspring of a schizophrenic parent will be affected is 13% compared with a population incidence of 1%. In one investigation, the frequency of schizophrenia in adopted offspring of schizophrenic parents was compared with the frequency of schizophrenia in a control group of adopted individuals of unaffected parents. In the 47 adopted children of schizophrenic mothers, five also suffered from the disorder. In contrast none of the 50 individuals in the control group were affected. Apparently there is little change in the rate of recurrence, regardless of whether a child is reared by its biological or adopted parent. Thus the reason why schizophrenia is familial is that family members share the same genes not that they share the same environment.

Twin studies

Twin studies provide a natural experiment with which to examine the relative contributions of nature and nurture. Monozygotic or identical twins are genetically identical; dizygotic twins share only 50% of their genes. Since both types are born at the same time, in the same family, the environment in which they grow up is likely to be similar. Of course, in practice there will be differences in the actual environment they experience. There may be differences in the position of the foetus, difficulties during birth, they may contract different diseases, their interaction with other family members, friends, teachers, etc. may be different, they will experience different life events such as accidents, and so on. It is also important to remember that some differences may come about simply through the action of chance and some perceived differences may be the result of error in experimental measurement. All these diverse effects are collected together under the umbrella term **non-shared environment**.

In certain cases, monozygotic twins are separated at birth and reared apart. While the number of such cases studied is much smaller than the bulk of twin studies, they provide an additional type of experiment where genetically identical individuals experience different environments. The results of this may be compared with the more common case where genetically identical individuals (monozygotic twins) experience similar environments.

The parameter usually measured in twin studies is **concordance**. This is defined as the percentage proportion of identical twins who both suffer from a disease or exhibit a phenotypic trait when that disease or trait occurs in one member of the pair. The difference in concordance between monozygotic and dizygotic twins is a measure of the extent to which genetic differences contribute towards the population variability of a trait, since the former are genetically identical, while the latter share 50% of their genes. This allows a parameter called **heritability** to be calculated. Heritability is the proportion of total population variability that can be accounted for by genetic variability. It is important to understand that heritability does not describe the frequency that a parental trait is inherited by their children. For example, if the heritability for height is said to be 50%, this does not mean that parents who are both six feet tall will have children six feet tall 50% of the time; rather it means that of the observed variation of height in a population, 50% may be ascribed to genetic variation. Heritability is commonly used to indicate the extent to which a trait is influenced genetically. As an example, the heritability of insulin-dependent diabetes is estimated to be 35%.

Table 1.9 shows that there is a strong genetic component to many common diseases. Clearly, lifetime health is strongly influenced by genetic constitution. It is interesting to note that diseases such as tuberculosis, which may have been assumed to be wholly environmental because it is caused by a bacterial infection, shows a high concordance between identical twins. This illustrates how resistance to infection is influenced by genetic factors. For most of the human race, for most of its history, infectious diseases were a common cause of death. It is hardly surprising that natural selection should operate to favour genes that increase resistance.

Table 1.9 Examples of twin concordance values for some common diseases.

Disorder	Concordance %	
	Monozygotic	Dizygotic
Breast cancer	6.5	5.5
Cleft lip	35	5
Insulin-dependent diabetes mellitus	50	5
Non-insulin-dependent diabetes mellitus	100	10
Multiple sclerosis	20	6
Peptic ulcer	64	44
Rheumatoid arthritis	50	8
Tuberculosis	51	22

Another interesting feature is that breast cancer shows relatively low concordance, despite the identification of two genes, *BRCA1* and *BRCA2*, which effectively result in simple autosomal dominant inheritance for the disease. The reason for this is that *BRCA1* and *BRCA2* only account for a small fraction of the total number of breast cancer cases.

1.3.2 Behavioural disorders and personality traits

Twin studies also show that some behavioural disorders such as schizophrenia have a genetic basis, at least in part (Table 1.10). One disorder that seems to be strongly influenced by genetic factors is autism. This disorder was previously considered to be entirely environmental, said to be caused by cold and aloof parents. Sparing such parents the additional trauma of such a label is a powerful benefit of such studies.

Table 1.10 Some examples for twin concordance values for common psychiatric disorders.

	Concordance %	
Disorder	Monozygotic	Dizygotic
Alcoholism	40	20
Autism	60	7
Schizophrenia	44	16
Alzheimer's disease	58	26
Major affective disorder	60	20
Reading disability (dyslexia)	64	40

Table 1.11 Twin concordance values for some fundamental human attributes.

	Concordance %	
Trait	Monozygotic	Dizygotic
General intelligence	80	56
Perceived happiness	50	8
Neuroticism	44	18
Extraversion	50	18

The application of twin and adoption studies to the study of normal phenotypic variation has revealed strong genetic components to many fundamental human attributes. Table 1.11 shows that twin concordance is high for general intelligence, personality (extraversion/neuroticism) and even perceived happiness. Most of the heritabilities are around 50%, so it is important to remember that this means that 50% of the variation is environmental.

The conclusions of these studies in behavioural genetics, if true, may have profound societal implications, which are discussed in Chapter 11. It is important to realise that the conclusions concerning the extent to which genetic factors control behaviour and personality are still controversial. There are four main areas of criticism.

1. The estimates of heritability require knowledge of the extent to which the genetic components are additive or epistatic (see Chapter 5). Procedures for analysing these components have been elaborated in model organisms where large numbers of progeny in controlled crosses can be studied. Clearly, in human inheritance such experiments cannot be carried out. The estimates of heritability are therefore correspondingly less rigorous, and lead to considerable debate about the methods of mathematical analysis used.

2. Early twins studies were based on small numbers. This has now been largely rectified by two large-scale studies, the Loehlin study and Minnesota twins reared apart (MISTRA). The latter assembled a dataset of 522 monozygotic twins and 408 dizygotic twins and established a registry of twins reared apart, which consisted of 59 monozygotic twins and 47 dizygotic twins.

3. It is difficult to precisely define personality traits and disorders. Such uncertainties are one reason why claims to have located genetic loci that contribute to personality disorders have subsequently been retracted (see Chapter 5). One major area of criticism has been systematic cultural bias in IQ tests.

4. Adoption studies and studies of twins reared apart have been criticised on the basis that the environments are not as different as might be supposed, because adoption agencies tend to match the ethnic, cultural and socioeconomic background of the biological and adoptive parents.

1.4 Chromosomal mutations

During gametogenesis the diploid complement of chromosomes is reduced during meiosis, so that each gamete receives one member of the 22 pairs of autosomes and one sex chromosome. During this process recombination takes place, resulting in the exchange of information on homologous chromosomes. Failure of this process results in gametes that do not have the correct complement of chromosomes; when such a gamete fuses with another the resulting zygote will have a chromosome imbalance. If the failure results in a whole chromosome failing to segregate, the event is called a **non-disjunction**. Non-disjunction can result in a gamete with an extra chromosome, in which case the zygote is said to be **trisomic**, or a gamete in which a chromosome may be missing, in which case the zygote will be **monosomic**.

Accidents at meiosis can also result in chromosome rearrangements, duplications and deletions. A reciprocal **translocation** is an exchange of segments between non-homologous chromosomes (Figure 1.5). If the pair of translocated chromosomes are inherited together there will be no change in the information content, apart from the breakpoint, which, if it occurs within a gene, may result in that gene being damaged. Such translocations have been very valuable in mapping disease genes, since the position of the breakpoint may be mapped cytogenetically by examining banded chromo-

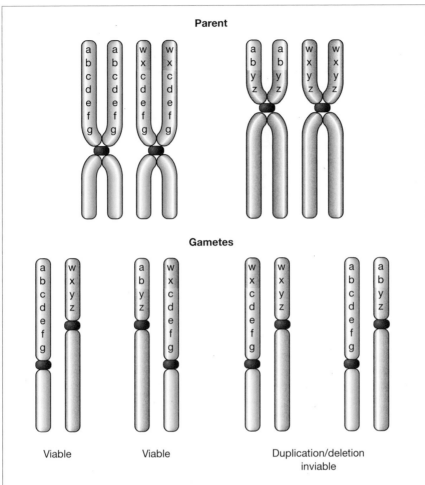

Figure 1.5 A balanced translocation between non-homologous chromosomes. Each gamete receives one copy of each non-homologous chromosome. Only two of the combinations will result in the normal gene complement. The other two combinations result in duplication and deletions of genes, resulting in inviability or a mutant phenotype.

somes during metaphase in an affected individual (see Chapter 4). If the pair of translocated chromosomes are not inherited together in a gamete, then in the resulting zygote some segments of chromosomes may be missing and some present in triplicate. Such gametes are not likely to be viable, so normally both members of a pair of reciprocally translocated chromosomes must be inherited together. Such a situation is described as a **balanced translocation** (Figure 1.5). An **unbalanced translocation** is the non-recip-rocal duplication of a chromosome segment (Figure 1.6). Inheritance of such a chromosome may lead to a chromosome imbalance, for example a minority of Down's syndrome cases are familial. They result from trisomy for part of chromosome 21 due to an unbalanced translocation. Deletions of chromosomes lead to monosomy of chromosomal regions. *Cri-du-chat*

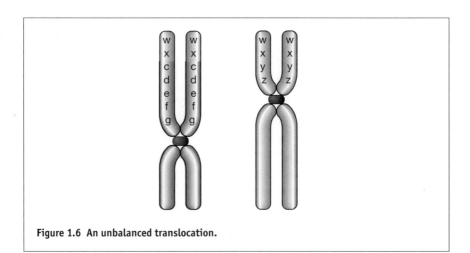

Figure 1.6 An unbalanced translocation.

syndrome is caused by a deletion of part of chromosome 5. This syndrome is characterised by a characteristic cat-like mewing and mental retardation.

Chromosomal mutations are very common. One survey of spontaneous miscarriages showed that in 50% the foetus was affected by a major chromosomal abnormality. Some foetuses carrying such mutations survive to term but are consequently affected by the chromosome imbalance. These form a major category of genetic disease. Some examples are shown in Table 1.12.

Table 1.12 Some examples of chromosomal mutations.

Condition	Frequency
Sex chromosomes	
45X: Turner's syndrome	1 in 5 000
47XXY: Klinefelter's syndrome	1 in 1 000
Autosomes	
Trisomy 21: Down's syndrome	1 in 800 (maternal age dependent)
Trisomy 18: Edward's syndrome	1 in 10 000
Balanced translocations	1 in 500

The most common is Down's syndrome caused by trisomy 21. The features of this syndrome are mental subnormality, characteristic broad facial features and short stature. This is usually due to an extra copy of the whole chromosome 21, although it can be caused by a partial trisomy due to a parent with a balanced translocation. The former result from accidents at meiosis and are therefore sporadic. Nevertheless, the frequency increases with maternal age. A young mother age 21 has only a 1 in 2000 chance of giving birth to a child affected by Down's syndrome, whereas the risk increases to 1 in 45 for mothers who are 45 years old. This is thought to be due to the fact that the mother's eggs start to develop before birth and become arrested at prophase of meiosis I. Thus, at the time of ovulation an egg may be over 45 years old and the consequent deterioration results in the decreased fidelity of the meiotic process.

Another very common class of chromosomal mutations are those involving sex chromosomes. Monosomy for the X chromosome (45X) results in Turner's syndrome, characterised by a sterile female phenotype with short stature and a web of skin between the neck and shoulders. Klinefelter's syndrome results from an XXY chromosome composition. This syndrome is characterised by a sterile male phenotype, tall and thin body form with breast development.

1.5 Mitochondrial mutations

Mitochondria provide 90% of cellular energy and thus the energy needed by organs, tissues and the body as a whole. Energy is generated in the mito- chondria by the respiratory chain in a process known as **oxidative phosphorylation**. The respiratory chain consists of five protein complexes (complexes I–V) involving a total of 90 separate proteins. During oxidative phosphorylation electrons are passed along the chain from one complex to another. At the same time protons are pumped out of the mitochondrial matrix, generating a gradient across the inner mitochondrial membrane. The protons flow back into the matrix through complex V (ATP) synthase, which provides the energy for ATP production.

Each cell contains hundreds of mitochondria. Each contains between two and ten copies of a 16.6-kb circular DNA genome (mtDNA; Figure 1.7). In terms of size this corresponds to 0.0006% of the nuclear genome, but because there are approximately 10 000 mtDNA molecules per cell it amounts to about 1% of the total mass of cellular DNA. mtDNA encodes a number of essential functions, which are translated within the mitochondria using a mitochondrial-specific protein synthesis apparatus. The functions encoded by mtDNA are as follows:

- 13 respiratory chain subunits
  - seven subunits of complex I (NADH dehydrogenase)
  - three subunits of complex IV (cytochrome *c* oxidase)
  - two subunits of complex V (ATP synthase)
  - cytochrome *b* (a subunit of complex III)

- tRNA for each amino acid

- 12S and 16S rRNA for mitochondrial ribosomes.

The remaining functions are encoded by nuclear genes and synthesised in the cytoplasm before import into mitochondria.

Impairment of mitochondrial function leads to a decline of energy availability and results in clinical disorders. In principle, mutations to genes in both the nuclear and mitochondrial genomes could bring this about. In the last 10 years mutations to mtDNA have been shown to be the cause of

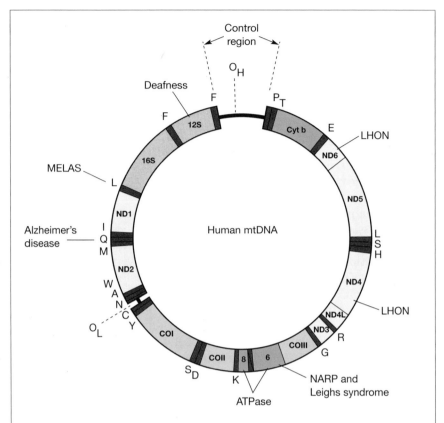

Figure 1.7 Map of the mitochondrial genome. The 22 tRNA genes are denoted by the single-letter amino acid code. 12S and 16S, 12S and 16S rRNA genes; COI–COIII, subunits of cytochrome *c* oxidase (complex IV); Cytb, cytochrome *b*; ND1–ND7, subunits of NADH dehydrogenase; ATP6 and ATP8, subunits of ATP synthase (complex V); O_H and O_L, origins of replication for the heavy and light strand respectively. The sequence of the control region is variable. It contains the D-loop, required for DNA replication.

a number of disorders. In general these are multisystem disorders affecting the central nervous system, sight, hearing, heart and skeletal muscles, the kidneys and endocrine glands (Table 1.13).

The part of the central nervous system most often affected is a region of the brain known as the basal ganglia. The basal ganglia is important for coordinated motion and its dysfunction results in ataxia. The effect of mitochondrial dysfunction on muscles is to cause mitochondrial myopathy, a general term describing the degeneration and loss of function of muscles associated with the presence of ragged red fibres: degenerating muscle fibres containing defective mitochondria that turn red in the presence of a specific stain. The exocrine gland most commonly affected is the pancreas, resulting in diabetes mellitus due to lack of insulin and pancreatic exocrine deficiency, which is a failure to secrete the digestive enzymes that originate in the pancreas.

Table 1.13 Diseases caused by mitochondrial mutations.

Disease	Acronym	Symptoms	Gene affected
Chronic progressive external ophthalomoplegia	CPEO	Mitochondrial myopathy and paralysis of eye muscles	Multiple gene loss due to deletion
Kearns–Sayre syndrome		CPEO plus ataxia, retinal deterioration, heart disease, hearing loss, diabetes and kidney failure	Multiple gene loss due to deletion
Leber's hereditary optic neuropathy	LHON	Blindness caused by damage to the optic nerve	Mutations in subunits of NADH dehydrogenase
Leigh's syndrome		Degeneration of basal ganglia leading to loss of motor and verbal skills	ATP synthase
Mitochondrial encephalomyopathy, lactic acidosis and stroke-like episodes	MELAS	Dysfunction of brain tissue causing dementia and seizures, mitochondrial myopathy and lactic acidosis	$tRNA^{Leu}$
Neurogenic muscle weakness, ataxia and retinitis pigmentosa	NARP	Muscle weakness, ataxia and blindness	Subunits of ATP synthase
Pearson's syndrome		Childhood bone marrow dysfunction, leading to multiple blood disorders and pancreatic failure	Multiple gene loss due to deletion

Mitochondria are maternally inherited, being passed on to progeny in the cytoplasm of the egg cell. Although the sperm acrosome contains mitochondria, they are not retained after fertilisation. Normally the 10 000 copies of the mitochondrial genome in each cell are identical and the cell is said to be **homoplasmic**. When a mutation occurs in one of the mitochondrial genomes there will be a mixture of different genome types in the cell and it is said to be **heteroplasmic**. The heteroplasmic state is short-lived and the descendants of a heteroplasmic individual become homoplasmic in a few generations. This is thought to occur because the oogonia contain a much smaller number of mitochondria (about 200) than other cell types, which creates a situation similar to a **population bottleneck** that reduces genetic diversity (see section 9.2.2).

Mitochondrial disorders do not show regular Mendelian patterns of inheritance. In fact the pattern of inheritance can be complex. There are a number of interrelated reasons for this.

1. Mitochondria are maternally inherited so mitochondrial disorders can only be inherited from an individual's mother.
2. If a mother is heteroplasmic each of her children may receive different proportions of affected mitochondria. Therefore there will be a wide variety in the type and severity of symptoms in each child.
3. During development and the lifetime of an individual derived from a heteroplasmic zygote, the proportion of defective genomes may change. This may occur through chance sampling effects during segregation of mitochondria to daughter cells after cell division, or through the acquisition of new mutations to previously wild-type genomes. As a result, in an affected individual, there will be variation in the proportion of defective genomes, both spatially in different tissues and temporally throughout the individual's lifetime. The effect of mutation will lead to an increase with time in the proportion of defective genomes. This is probably the reason why mitochondrial disorders often only become manifest after a delay of several years and why some disorders become progressively more severe with age. A factor that will exacerbate this process is that interruption to the electron transport chain leads to the production of highly reactive free oxygen radicals, which are mutagenic. Thus a slight impairment to mitochondrial function, which initially is not severe enough to cause clinical symptoms, may nevertheless interrupt electron transport sufficiently to trigger this process and eventually result in the onset of a disorder.

Both large-scale deletions and mutations to specific genes can affect the mitochondrial genome.

Point mutations affecting NADH dehydrogenase cause LHON. This disease predominantly affects young men, resulting in blindness due to damage to the optic nerve and sometimes heart and neurological abnormalities.

Mutations to ATP synthase result in variable phenotypes. Severely affected individuals suffer from Leigh's syndrome, a devastating and often lethal childhood disorder in which the basal ganglia of the brain degenerate. Less severely affected individuals suffer from NARP, characterised by muscle weakness, ataxia and a form of blindness called retinitis pigmentosa (which can also be caused by a variety of chromosomal mutations). The difference in severity in different individuals with the same mutation reflects the proportion and tissue distribution of mutant mtDNAs in affected individuals. Individuals with Leigh's syndrome have high levels of mutant mtDNA in multiple tissues.

Point mutations affecting tRNA or rRNA have severe effects because they simultaneously reduce the ability to make different mitochondrial proteins. MELAS is caused by a mutation in the tRNA^{Leu} gene. Other mutations of this gene have been shown to result in a variety of severe symptoms, including mitochondrial myopathy, heart disease and diabetes mellitus. Indeed, mutations to tRNA^{Leu} are responsible for 1.5% of all incidences of diabetes mellitus. Mutations to tRNA^{Glu} have been found to be associated with 5% of patients with late-onset Alzheimer's disease. Mutations in the 12S rRNA gene also generally affect protein synthesis. They have been shown to be responsible for congenital deafness.

The effect of deletion in mtDNA (Δ mtDNA) is different in adults and children. Adults generally suffer from slowly progressive neurodegenerative disorders such as CPEO and Kearns–Sayre syndrome. Children suffer from severe disorders, such as Pearson's syndrome, that progress rapidly and affect multiple organs. CPEO is characterised by mitochondrial myopathy and paralysis of the eye muscles. Kearns–Sayre syndrome is characterised by CPEO plus retinal degeneration, heart disease, ataxia, deafness, diabetes mellitus and kidney failure. Pearson's syndrome causes a failure to make blood cells, resulting in severe anaemia and other blood disorders, pancreatic failure leading to diabetes mellitus and pancreatic exocrine dysfunction. Children who suffer from Pearson's syndrome generally die in the first few years of life. A few children who do survive generally develop Kearns–Sayre syndrome as they grow older.

The differences in clinical phenotypes between children and adults may result from differences in the proportion and tissue distributions of Δ mtDNA. A more widespread distribution with a higher proportion of Δ mtDNA will result in earlier onset with more severe symptoms. The tendency of mitochondrial deletions to worsen over time is probably caused by preferential replication of Δ mtDNA in non-dividing cells. The reasons for this are not clear at present. The origin of the deletions is also unclear, since they are rarely passed on from a mother with Δ mtDNA, presumably because an egg cell containing Δ mtDNAs could not survive. mtDNA with duplicated segments is often found to be associated with Δ mtDNA. A duplicated genome would not result in any impairment since all functions are still present. However, they may give rise to deletions as a result of complex intramolecular recombination events between the duplicated segments.

Mitochondria and ageing

Common symptoms of mitochondrial disorders, such as diabetes, dementia, muscle weakness, loss of sight and hearing, ataxia, etc., are also characteristic of the normal ageing process. This has led to the suggestion that accumulation of somatic mutations in mtDNA may contribute to the deterioration of body function with age. Many environmental toxins inhibit mitochondria and so may be a contributory factor. Evidence is accumulating in support of this hypothesis.

- The performance of the respiratory chain complex deteriorates with age in the brain, skeletal muscle, heart and liver.

- Rearrangements in mtDNA accumulate with age in these same tissues. Δ mtDNAs accumulate in skeletal muscle after age 40, consistent with other observations that show that Δ mtDNA is preferentially replicated in non-dividing tissue.

- Genetic predisposition to type II diabetes often tends to be maternally inherited, consistent with a mitochondrial component to the aetiology.

- Animals raised on restricted-calorie diets have longer lifespans. Animals fed on such diets produce fewer free oxygen radicals and accumulate less damage to mtDNA.

- Individuals who suffer from ischaemic heart disease experience temporary interruptions to the blood flow to the heart due to atherosclerotic plaques. During the resultant anoxia in the heart muscle, the respiratory chain is blocked; upon restoration of the blood supply (reperfusion) there is a burst of free oxygen radical production. The mitochondria in heart muscle of such patients contain a high level of mutant mtDNA, which may accelerate the onset of heart failure.

One intriguing possibility is that dietary supplements of antioxidants such as vitamin C may help limit the production of free radicals in the mitochondria and thus delay the ageing process.

1.6 Summary

- Genetic factors play a major role in determining lifetime health. Approximately 5% of babies born alive will suffer from a significant medical condition for which there is a genetic component. Susceptibility to many of the common diseases of later life is also influenced by our genetic constitution.

- There are four main classes of genetic diseases: (i) monogenic defects; (ii) multifactorial or complex disorders; (iii) chromosomal imbalances; and (iv) mitochondrial mutations.

- Monogenic disorders follow Mendelian patterns of inheritance, and are subdivided into autosomal recessive, autosomal dominant and sex linked.

- In monogenic disorders, the pattern of inheritance and the manifestation of the symptoms may be complicated by a variety of factors, such as genetic heterogeneity, penetrance, varying expressivity, mosaicism, anticipation, genomic imprinting and phenocopy.

- Multifactorial or complex disorders come about through the interaction between polygenic or oligogenic determinants and the environment. Evidence for the genetic component comes from family, adoption and twin studies.

- Research in behavioural genetics suggests that common psychiatric disorders such as schizophrenia may have a strong genetic determinant in their causation. This type of research also suggests that many fundamental human attributes have heritable components. Such research is still highly controversial.

- Chromosomal mutations arise through non-disjunction at meiosis, producing aneuploids. An extra copy of a chromosome results in trisomy, while loss of a chromosome results in monosomy. Trisomy 21 is responsible for Down's syndrome, one of the commonest genetic conditions. Other common chromosome abnormalities are balanced and unbalanced translocations, and deletions.

- Mitochondrial disorders are maternally inherited. A variety of factors lead to great complexity in the age of onset and severity of symptoms. Mitochondrial disorders affect many different organs simultaneously, including the central nervous system, skeletal muscle, heart, sight, hearing and endocrine function. Such disorders tend to become progressively worse as further mitochondrial mutations accumulate.

Further reading

General

House of Commons Science and Technology Committee Third Report (1993) *Human Genetics: the Science and its Consequences. Volume 1. Report and Minutes of Proceedings*. HMSO, London.

Provides a general introduction to the ground covered by this book. As well as giving data for the incidence of various diseases, it provides an account of the views of leading clinicians, scientists and industrialists as to the influence of genetics on human health and the possibilities of the new developments to ameliorate the consequences of genetic defects. Finally, it provides a commentary on the legal, ethical and social impact of human genetics along with recommendations designed to ensure that scientific developments do not have a negative impact, issues discussed in Chapter 11 of this book.

WEATHERALL, D.J. (1991) *The New Genetics and Clinical Practice*, 3rd edn. Oxford Medical Publications, Oxford.

Chapter 1 provides a detailed review of the influence of genetic factors on human health. It is notable for the carefully researched data on the frequency of different genetic diseases and the extent of ethnic variation.

Monogenic disorders

Online Mendelian Inheritance in Man (OMIM)
http://www3.ncbi.nlm.nih. gov/omim

This database started life as a catalogue assembled by Victor McCusick, one of the leading figures in the field of human genetics. It is now a Web site maintained by the National Center for Biotechnology Information (NCBI) in the United States. It contains a total of 5605 gene loci (as of August 1997) that are different from each other and where the mode of inheritance is judged to have been proved. Each entry contains a textual review of the locus and, where available, links to pictures.

Complex diseases

LANDER, E.S. and SCHORK, N.J. (1994) Genetic dissection of complex traits. *Science*, **265**, 2037–2048.

GHOSH, S. and COLLINS, F.S. (1996) The geneticists approach to complex disease. *Annual Review of Medicine*, **47**, 333–353.

Inheritance of psychiatric disorders and personality traits

BOUCHARD, T.J. (1994) Genes, environment and personality. *Science*, **264**, 1700–1701.

MANN, C.C. (1994) Behavioural traits in transition. *Science*, **264**, 1686–1689.

PLOMIN, R., OWEN, M.J and McGUFFIN, P. (1994) The genetic basis of complex human behaviours. *Science*, **264**, 1733–1739.

ROSE, S.J., LEWONTIN, R.C. and KAMIN, L.J. (1990) *Not in our Genes: Biology, Ideology and Human Nature*. Penguin, Harmondsworth.

An outspoken critique of the evidence that underpins the claims that genetic factors control behaviour and the incidence of psychiatric disorders.

Mitochondrial disorders

LARSSON, N.G. and CLAYTON, D.A. (1995) Molecular aspects of human mitochondrial disorders. *Annual Review of Genetics*, **29**, 151–178.

WALLACE, D.C. (1997) Mitochondrial DNA in aging and disease. *Scientific American*, **276** (8), 22–29

CHAPTER TWO

Structure of the human genome

Key topics

- Amount of DNA in the human genome
- Genes
  - ○ Gene expression
  - ○ Gene families
  - ○ Pseudogenes
- Tandem repeat arrays of rRNA, tRNA and histone genes
- Intermediate repeated DNA
  - ○ LINEs: L1 and retrotransposition
  - ○ SINEs: *Alu* family
  - ○ Processed pseudogenes
  - ○ Selfish DNA
- Highly repetitive DNA
  - ○ Telomeres
  - ○ Tandem repeat arrays at centromeres
  - ○ Minisatellites
  - ○ Microsatellites
- Human karyotype
- Packaging DNA into chromosomes
  - ○ Nucleosomes
  - ○ 30-nm fibre
  - ○ 700-nm fibre

2.1 Introduction

The human **genome** is the term used to describe the sum total of DNA molecules found within every cell of the human body except red blood cells. Nearly all the genome is found in the nucleus; however, the mitochondria

found in a gamete, is half the complement of a somatic cell. The human
genome consists of 3000 million base pairs of DNA (3×10^9 bp or 3000 Mb).

This DNA contains all the information required for a zygote to develop
into an adult human. The information is in the form of genes. Each gene
consists of a length of DNA that performs some function, usually specifying
the amino acid sequence of a protein. A surprising aspect of the human
genome is that genes form only 3% of the total. Some of the rest of the
DNA is non-coding, but has a functional role in regulating and promoting
gene expression or a structural role in chromosome integrity and segregation of chromosomes at nuclear division. A large fraction of the genome
does not appear to have a function; if it does, it is a function that does not
depend on the sequence of bases within the DNA. Much of this DNA consists of a menagerie of repetitive and mobile elements that may be parasitic
or 'selfish' in origin. The genes themselves are not continuous stretches of
DNA: the **coding sequences (exons)** are interrupted by non-coding
sequences (**introns**). Sometimes the exons form only small patches in long
stretches of introns. Indeed, as discussed below, in some cases only 1% or
less of the total DNA that forms a gene may be coding sequence.

Thus the human genome is large, and complicated in its organisation.
Before gene cloning techniques were developed, it was impossible to study
individual genes. Even with gene cloning, it is still often very difficult to
find and clone human genes. The small fraction of the genome that forms
genes and the technical complications introduced by the repetitive and
mobile elements exacerbates the problem further.

The size and complexity of the human genome are a constant backdrop
to the whole field of human molecular genetics. Special techniques have
been developed to overcome these problems. This book is an account of
these techniques and how their application has led to spectacular progress
in recent years towards understanding the nature of human genes and how
they influence our health. Before this account can begin, it is necessary to
describe the structure of the genome and how the DNA is packaged into
chromosomes. This account is only intended to be a summary of material
treated in much more detail elsewhere. You should also refer to one of the
several excellent textbooks listed at the end of the chapter.

2.2 Sequence architecture of the human genome

In the 1960s a simple experiment measuring the rate of reassociation of
denatured DNA led to the surprising conclusion that a large fraction of
mammalian DNA consisted of repetitive sequences (see Further reading). As
a result of these experiments, human DNA may be categorised as follows.

1. **Single sequence DNA** or low-copy DNA (about 60% of the total). This
   class contains the sequences that form genes, as would be expected
   from the Mendelian laws of genetics. However, coding sequences only

account for about 3% of the total genome. The remainder of this class is DNA from introns or from spacer DNA, sequences of no apparent function that separate genes.

2. **Intermediate repeated DNA** present at between 10^2 and 10^5 copies per genome (about 30% of the total).

3. **Highly repetitive DNA** present at over 10^6 copies per genome (about 10% of the total).

2.2.1 Genes

A functional definition of a **gene** is a DNA sequence that contributes to the phenotype of an organism in a way that depends on its sequence. A change in the sequence may affect its function and may have phenotypic consequences. Most genes encode proteins. However, sequences that encode functional RNA molecules are also classified as genes, as their function depends on their sequence.

Gene expression

Transcription in humans, as in all other eukaryote cells, is catalysed by three types of RNA polymerase, designated RNA polymerase I, II and III (or pol I, pol II and pol III respectively). RNA polymerase I transcribes rRNA. RNA polymerase II transcribes nuclear genes that encode proteins. RNA polymerase III transcribes tRNA genes, and a small number of functional RNA molecules such as 5S RNA and nuclear RNAs that mediate splicing (see below). All three RNA polymerases require the action of transcription factors to bind DNA and promote transcription.

The structure of the **upstream** region of a typical gene transcribed by pol II is shown in Figure 2.1. A typical gene contains a **promoter** region, which extends for about 200 bp upstream of the transcription start site at nucleotide +1. In addition it may contain one or more of the following elements: **enhancers**, **response elements** and **silencers**. The promoter contains all of the elements required for a basal level of transcription. These elements are short conserved sequences that bind different **transcription factors**. Transcription factors are said to be *trans*-acting because they are encoded by another gene that must be translated in the cytoplasm and imported into the nucleus. The sequence elements on the DNA to which they bind are said to be *cis*-acting because they control the transcription of a gene that lies immediately adjacent on the same DNA molecule.

Most human promoters contain an element called a TATA box centred around position –30 relative to the transcription start site. The TATA box binds **general transcription factors**, designated TFIIX, where the roman numeral denotes the RNA polymerase involved and X denotes the particular protein. In pol II transcription the process is initiated by the binding of TFIID mediated by another protein called **TATA-binding protein** (TBP). This is followed by the assembly of an initiation complex with pol II and the other general transcription factors.

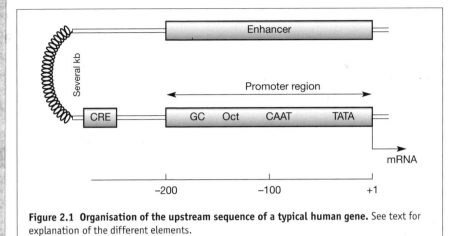

Figure 2.1 Organisation of the upstream sequence of a typical human gene. See text for explanation of the different elements.

The TATA box is responsible for correctly locating the site at which transcription starts. Three other commonly found elements within the promoter affect the efficiency of transcription.

- **GC box** (GGGCGG), which binds the transcription factor SP1.

- **CAAT box** (GGCCAATCT), which binds the transcription factors CTF and NF1.

- **Oct box** (ATTTGCAT), which binds the transcription factors Oct-1 and Oct-2.

The CAAT box is normally found around position –80 relative to the transcription start site. The spacing of the GC box and Oct box may vary. Active promoters contain at least one of these elements, but do not necessarily contain all three. As well as these elements that bind ubiquitous transcription factors, tissue-specific elements may also be found within the promoter region.

Enhancers are regions of DNA that operate to stimulate the basal level of transcription from its promoter. They are operationally distinguished from promoters by three criteria:

- they may be located a considerable distance from the transcription start site;

- their action is not dependent on their location: they may be located **upstream** or **downstream** of the gene they control;

- their action is not dependent on orientation.

Enhancers usually contain multiple sites where transcription factors bind to stimulate transcription. These may include sites for the same ubiquitous transcription factors found within the promoter as well as sites for the binding of specific transcription factors. Enhancers probably act at a distance through the formation of DNA loops, which bring the transcription factors bound to the enhancer region into close proximity with the basal transcription complex.

Response elements induce genes in response to particular signals. Examples include cyclic AMP response element (CRE), serum response element (SRE) and heat-shock response element (HRE). Response elements may be found within the promoter region, closely upstream of the promoter or in more distant enhancers.

Transcription control is generally positive in eukaryotes: transcription is stimulated by the presence of transcription factors. However, there are examples of negative control where genes are turned off by the binding of proteins to elements known as **silencers**. In yeast, where this phenomenon has been well documented, the same transcription factor can stimulate transcription at one gene and silence transcription at another. The action of silencers is less well characterised in humans.

The overall level of transcription of a gene is the outcome of the different influences exerted by the promoter and enhancers. Some genes, termed **housekeeping genes**, are expressed in most tissues most of the time, and are responsible for functions likely to be necessary in any cell. Such genes may be transcribed from a promoter alone if it contains sufficient elements for the binding of ubiquitous transcription factors such as SPI1. Other genes, whose transcription is tissue specific or depends on the presence of a specific signal, require the action of enhancers and response elements to stimulate the basal transcription from their promoters. There is great scope for sophisticated fine tuning of transcription arising from the interaction of many transcription factors to sites within the promoter and enhancers.

As soon as transcription starts, the 5' end of the nascent mRNA molecule is covalently modified to form a structure known as a **cap**. This is formed by the addition of 7-methylguanosine to the 5' nucleotide of the nascent mRNA in a 5'-5' triphosphate linkage. This effectively blocks the 5' end of the molecule since the 7-methylguanosine is in the opposite orientation to the rest of the molecule. The cap is essential for initiation of translation at the first AUG codon in the mRNA. The appearance of sequence AAUAAA in the nascent RNA molecule causes transcription to terminate a few hundred base pairs downstream; the RNA molecule is then cleaved about 20 bp downstream of the AAUAAA signal. The 3' end is also covalently modified by the addition of 100–200 adenine residues to form a structure known as the poly-A tail.

The part of the gene that encodes the protein contains sequences called introns which interrupt the protein-coding regions, known as **exons** (see Figure 2.2). The **primary RNA transcript** contains sequences derived from the introns. These are removed by a process known as **splicing** before the transcript leaves the nucleus. Splicing occurs through a cyclical process that occurs in a structure called a spliceosome. The spliceosome consists of small nuclear RNA molecules complexed to proteins called snRNPs (small nuclear ribonucleoprotein particles). These are sometimes colloquially known as 'snurps'. Splicing requires the concerted action of snRNPs and specific sequences at the junctions of introns and exons called **splice donor** and **splice acceptor** sites.

The splice donor site is sometimes known as the 5' splice site; it consists of the first two nucleotides of the intron, which is always the dinucleotide GT. The splice acceptor site is sometimes known as the 3' acceptor site; it consists of the last two nucleotides of the intron, which is always the dinucleotide AG. Alteration to these sequences will prevent splicing and will have phenotypic consequences. In addition, there is a sequence in the intron known as the branch site, 18–40 nucleotides upstream of the splice acceptor site, where a splicing intermediate called a **lariat** is formed. However, in humans this sequence is not highly conserved and, if altered by mutation, splicing can proceed using other related sequences in the vicinity.

RNA processing is a point at which gene expression can be modified. There are now many examples of genes that are differentially spliced and give rise to sets of different polypeptides that originate from the same primary transcript.

CpG islands
DNA can be modified by methylation of cytosine to form 5-methylcytosine. This normally occurs at the dinucleotide CpG, i.e. a C residue followed in the 3' direction by a G residue on the same strand of DNA. The dinucleotide CpG occurs less frequently than would be expected by chance from the base composition of DNA. This is a consequence of DNA repair mechanisms. Accidental deamination of cytosine produces uracil, whereas deamination of 5-methylcytosine produces thymine. Because uracil is not found in DNA, it is efficiently excised and replaced with cytosine by an enzyme called glycolase. However, thymine, a natural constituent of DNA, is now mispaired with a G residue. Although this will also be repaired, the process is not perfectly efficient and over a long period has led to a gradual reduction in the frequency of the CpG dinucleotide. DNA methylation is generally associated with repression of transcription.

So CpG is relatively rare in the human genome and where it occurs it will normally be methylated. However, so-called **CpG islands** exist where the frequency of CpG is greatly elevated, approaching that expected from the percentage (G+C) base composition of the human genome, and the CpG dinucleotides are hypomethylated or not methylated at all. These CpG islands are about 1 kb in length and tend to extend over promoters of expressed genes; about 56% of human genes are estimated to be associated with such sequences. They can be recognised by restriction enzymes that contain CpG in their target sequence and which will thus cut less frequently than expected because of the relative rarity of the CpG dinucleotide and because methylation inhibits the action of the restriction enzyme. Sites where these enzymes cut DNA will be clustered in CpG islands.

Size, spacing and number of human genes
A question that has provoked much interest over the years is how many genes there are in total in the human genome. There have been a number of approaches to answering this question.

- Theoretical calculation of the number of mutant genes that could be tolerated and the observed mutation of individual human genes places a ceiling of about 100 000 genes in the human genome. The calculations require many questionable assumptions, such as extrapolating the average mutation rate of all genes from the observed rate of a few.

- The total number of CpG islands is estimated to be 45 000. If 56% of genes have CpG islands the total number of genes is estimated at 80 000.

- The number of different expressed sequence tags (see Chapter 3) collectively present in human tissues is estimated to be about 65 000. This provides a measure of the total number of sequences expressed as mRNA.

- Extrapolation from the observed density of human genes in areas already well characterised leads to a figure of 70 000.

Completion of the Human Genome Project will provide a definitive answer to this question (see Chapter 8). In the mean time, independent methods provide reasonable agreement that there are between 65 000 and 80 000 genes in the human genome.

Human genes show enormous variation in overall size and the size and number of introns. Some genes, such as the histone genes, do not contain introns. The α-globin gene is 0.8 kb in size and contains three introns which account for 30% of the total genomic locus. Other genes are extremely large. The dystrophin gene (the gene defective in DMD), is 2.4 Mb in size and contains 79 introns, which account for 99.4% of the total genomic locus. Other examples of large genes are those for factor VIII (haemophilia A), which is 186 kb in size, and CFTR (cystic fibrosis), which is 250 kb in size. In each case introns account for over 95% of the genomic locus. The structures of the α-globin and factor VIII genes are compared in Figure 2.2.

Sometimes, then, human genes appear to consist of small islands of exons in large oceans of introns. This is an important point to bear in mind in the analysis of data in the Human Genome Project (see Chapter 8). Given this enormous range of size, it is not easy to give an average figure for the size of human genes but it is probably about 10–15 kb. Nor is it easy to give an average figure for the spacing of human genes because some regions are relatively gene-rich and others gene-poor. A rough estimate is about one gene per 45 kb DNA. Again, accurate figures will be available on completion of the Human Genome Project.

Compiling the information on gene number, size and spacing we can estimate that about 3% of the human genome actually encodes proteins (i.e. are exons). The genes that contain these protein-coding sequences probably occupy about 25–35% of the total genome.

Gene families
Between 25 and 50% of protein-coding sequences in the genome are unique. The remainder belong to families of similar or related genes. Some of these genes are dispersed through the genome, while others are present in clusters of related genes.

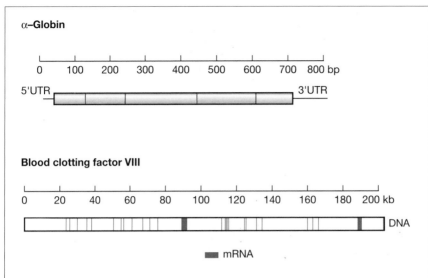

Figure 2.2 Human genes vary in size and intron content. The figure compares the size and intron content of the genes for α-globin and blood clotting factor VIII. In each case the coding regions are shown in solid blue boxes and the introns as open boxes. Note the difference in the scale shown above each gene. In the case of α-globin the 5' and 3' untranslated regions of the mRNA are shown (5' UTR and 3' UTR); these are normally classed as exons. The blood clotting factor VIII gene occupies 186 kb of the human X chromosome. It contains 26 exons ranging in size from 69 to 3106 bp and introns as large as 32 kb. The mRNA is only 9 kb in size and is shown to scale. It comprises a 7053-nucleotide coding sequence and a 3' UTR of 1806 nucleotides.

Gene families are thought to have arisen by a process of gene duplication and divergence from an ancestral gene. Figure 2.3 shows how the evolution of the different proteins that make up haemoglobin is thought to have occurred. Haemoglobin is the oxygen-carrying molecule in blood, comprising two β-globin and two α-globin protein subunits, each complexed to a molecule of haem. The different members of the globin gene families are expressed at different times during foetal, embryonic and adult stages of development (see Chapter 4).

As well as whole genes it is very common to find proteins of different overall function containing similar protein **domains**. For example, many different proteins contain nucleotide-binding domains but show no similarity throughout the rest of the protein.

Because there are multiple copies of genes in a family, sometimes loss of function in one of them may be tolerated. This has led to the evolution of **pseudogenes**, which are sequences recognisably similar to functional genes but which have accumulated nonsense and frameshift mutations that will prevent them from functioning. They may be viewed as the decaying relics of genes that were once functional. For example, the β-globin gene cluster contains two pseudogenes (Figure 2.3).

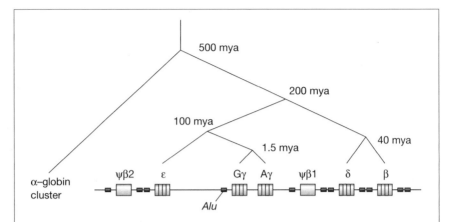

Figure 2.3 Evolution and structure of the β-globin gene family. About 600 million years ago (mya) gene duplication of a single globin-like gene produced the two genes that diverged to give rise to the ancestral genes of the α- and β-globin gene clusters. Initially the α and β genes were found as tandem repeats, as is still the case in modern amphibians. A translocation event separated the two genes to different chromosomes and this was followed by further rounds of gene duplication and divergence. ψβ1 and ψβ2 are pseudogenes. The positions of eleven *Alu* elements are shown by the solid boxes.

Non-protein-coding genes

Not all genes encode proteins. The bulk of cellular RNA consists of rRNA and tRNA required for protein translation. There are also a number of other RNA molecules that function directly within the cell. The 7SL RNA molecule forms part of the signal recognition particle required for the translocation of proteins across the endoplasmic reticulum. This is the first step in the secretory pathway that will result in the protein being exported from the cell. Small nuclear RNA molecules form complexes with protein to mediate the splicing of protein-coding RNA transcripts (see above).

Tandem repeat arrays

The rRNA molecules consist of three separate species: 28S, 18S and 5.8S. They are transcribed from a single transcription unit by a dedicated RNA polymerase known as RNA polymerase I. A single 45S transcript is produced that is processed to form the separate species. tRNA genes are also transcribed by a dedicated polymerase known as RNA polymerase III. (Protein-coding genes are transcribed by RNA polymerase II.)

The transcription units for rRNA are repeated several hundred times, so that each copy lies immediately adjacent to the next and are arranged so that the end of one unit abuts the start of the next. Such an arrangement is known as a **tandem repeat array** (Figure 2.4). The genes encoding the five histone proteins (see below) are also found in similar tandem repeat arrays. In contrast to gene families, the sequences of these genes in tandem repeat arrays are all identical or nearly identical.

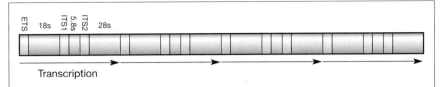

Figure 2.4 Tandem repeat array rRNA transcription units. Each unit is transcribed to produce a 45S transcript, which contains the ETS (external transcribed spacer unit), 18S, ITS1 (internal transcribed spacer 1), 5.8S, ITS2 and 28S sequences. This transcript is processed in a number of sequential steps to release the 18S, 5.8S and 28S RNA molecules.

rRNA and tRNA constitute about 90% of the total RNA within the cell. Histones are also present in large quantity as they are bound to DNA in a stoichiometric fashion (see below). When the DNA is replicated the amount of histones must also double, so large quantities have to be synthesised in a short period of the cell cycle. A single copy of the histone, rRNA and tRNA genes would not be capable of synthesising enough product. Only by having several hundred copies can the rate of synthesis keep up with demand.

Spacer regions

Genes are separated from each other by long tracts of DNA known as **spacer regions**. The sequence of spacer regions changes much more rapidly in evolution than the sequence of genes, indicating that there is less selection against changes in its sequence. Thus it is thought that spacer DNA does not have a function that depends on its sequence. It is possible that the physical separation of genes is important in some way, so spacer DNA may have a sequence-independent function. The long tracts of DNA found in introns may be regarded as a particular form of spacer DNA. An example of the physical separation of genes may be seen in the region that contains the five members of the β-globin gene cluster. Each gene is 1.6 kb in size including two introns, yet the cluster is spread over 60 kb of chromosome 11 (Figure 2.3).

2.2.2 Intermediate repeated sequences

Much of the intermediate class of DNA consists of a small number of sequence families. Each family consists of elements that have similar but not identical sequences. Most of these are **mobile genetic elements**, so called because new copies are generated in a new location, by a process of transposition. When this happens, one copy is found at the original location and one copy at the new location, thus duplicating the original DNA sequence. Since the removal of the extra copies is a slow process on an evolutionary time scale, they will increase in number. In humans they have come to occupy as much as 30% of the genome. This has led to the concept of **selfish** or **parasitic DNA** because they do not contribute to the phenotype of the organism but evolve only to increase in number. The signature of mobile genetic elements is a short direct repeat sequence either side of the point of insertion into the host chromosome.

There are two main types of mobile elements: **long interspersed elements** (LINEs) and **short interspersed elements** (SINEs). The most common and best-characterised LINE is called **L1** (Figure 2.5). L1 is repeated about 50 000 times in the human genome, accounting for about 5% of the total. Only 3 000 of these are full length; the remainder are truncated, mostly at the 5' end. The complete element is 6 kb in size and contains two sequences that could encode proteins (**open reading frames or ORFs**). One of the ORFs encodes a protein with similarities to **reverse transcriptase** found in **retroviruses**. An AT-rich region is located near the 3' end of the element and flanking the element are two short direct repeats. Three characteristics of this organisation suggest that L1 is a type of mobile genetic element called a **retrotransposon**.

1. Reverse transcriptase is known to mediate the transposition of other types of mobile genetic elements found in other organisms, such as the TY1 element in yeast.

2. There is a poly-A sequence at the end of each monomer that resembles the poly-A tail of an mRNA molecule.

3. The short direct repeats at either side of the element are characteristic of all mobile elements.

Retrotransposons are thought to have arisen when an mRNA molecule is copied into DNA by reverse transcriptase. The copied DNA then integrates into the DNA at a new site. Some pseudogenes lack introns present in their functional counterparts. These may also have resulted from **retrotransposition**. The mRNA derived from the original gene is processed to remove introns. This becomes the template for reverse transcriptase, resulting in a DNA copy of the gene lacking introns. This can then reinsert itself into the genome at a new location. Such elements are known as **processed pseudogenes**.

The main type of SINE is called the ***Alu*** family, because it commonly contains a target for the restriction enzyme *Alu* I. Members of the *Alu*

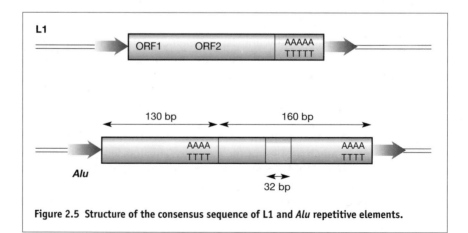

Figure 2.5 Structure of the consensus sequence of L1 and *Alu* repetitive elements.

is about 290 bp and consists of two tandem repeats of a 130-bp sequence. However, one repeat contains a 32-bp insertion (Figure 2.5). Like L1, *Alu* elements are thought to have arisen by retrotransposition:

1. they are flanked by direct repeats;
2. each repeat unit has an AT-rich region that suggests an intermediate with a poly-A tail;
3. the 5' end resembles a pol III promoter region.

7SL RNA (see above) was probably the progenitor, since each repeat unit of the *Alu* element is similar to part of the sequence of this gene. They evolved about 65 million years ago. There are between 5×10^5 and 10^6 copies of the *Alu* element in the haploid genome, forming about 1% of the total. On average there is one repeat every 4 kb. The 60-kb β-globin region contains 11 *Alu* elements. *Alu* elements are not found within coding sequences but are often present in the transcription unit within introns and occasionally in non-translated regions of the mRNA.

Although mobile elements do not contribute directly to the phenotype, they may still affect it. The insertion of a mobile element into a gene may result in its inactivation. Recent evidence suggests that *Alu* elements can integrate into control regions and remain there unmodified for millions of years. These elements bind regulatory proteins and assist in the regulation of transcription. They may also have played an accidental role in other forms of evolution. For example, unequal crossing-over between repetitive elements may be the cause of the gene duplication from which gene families arose (Figure 2.6).

2.2.3 Highly repetitive DNA

Highly repetitive DNA consists of short sequences repeated up to a million times in the genome. Because the average base composition of the repeated sequence may be different from the average of the rest of the genome, it often has a different buoyant density. Thus when total human DNA is fractionated by **bouyant density ultracentrifugation**, the highly repetitive DNA may form separate or satellite bands to the main peak. For this reason highly repetitive sequences are often known as **satellite DNA**.

Most satellite DNA is found in very long tandem arrays in heterochromatic regions around the centromere. **Heterochromatin** is the term used to describe chromosomal regions that are transcriptionally inactive and stain more densely with **Feulgen**, because it is condensed even in interphase. It also characteristically replicates later in the cell cycle than **euchromatin**. An important example of a satellite DNA sequence is α-satellite DNA. The α satellite is a highly repetitive sequence that has an essential function in the **centromeres** of chromosomes (Box 2.1). Centromeres are the point where chromosomes attach to the spindle at nuclear division, ensuring the proper segregation of chromosomes to daughter cells

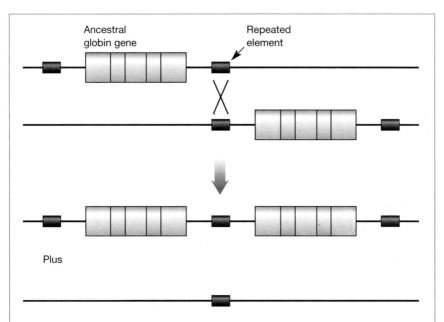

Figure 2.6 Gene duplication caused by unequal crossing-over. At meiosis, two homologous chromosomes are misaligned. Recombination between two repeated sequences results in a chromosome with a gene duplication and a chromosome with a deletion. The latter is presumably inviable and is lost.

BOX 2.1: α SATELLITE DNA

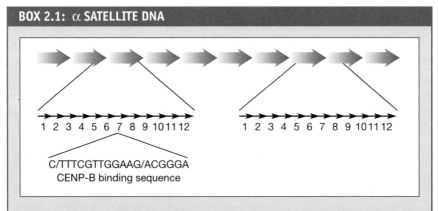

Each centromere contains a tandem array of α satellite repeats that extend for millions of base pairs, uninterrupted by any other sequence. Each array is organised in a hierarchy of higher-order repeats. These higher-order repeats are represented by the large arrows in the figure. Each higher-order repeat contains a number of monomers, which varies between 4 and 32 depending on the particular chromosome; these are represented by the smaller arrows numbered 1–12 in the figure. Some, but not all, of these monomers contain a 17-bp binding site for the centromere-specific DNA-binding protein CENP-B. The monomers are not identical to each other; indeed they can show up to 20% sequence divergence. They may be no more similar to each other than they are to the α satellite in other primates. In any one array, each of the higher-order repeats (large arrows) are virtually identical (<2% sequence divergence). They contain the same monomer subunits repeated in the same order. The organisation and sequence of the monomers is particular to each chromosome. The number of higher-order repeats varies from 100 to 5000 on different chromosomes, giving a range of overall array sizes from 0.2 to 10 Mb. Altogether, the α satellites represent several per cent of the total genome.

The repeat structure of the α-satellite sequences makes them extremely difficult to clone in bacteria, since they are continually rearranged by recombination. Nevertheless this has recently been achieved, and it has been shown that when introduced into cells in culture they will function as centromeres. They have been used to construct artificial human chromosomes.

Minisatellites and microsatellites

Some types of satellite DNA are interspersed throughout the whole of the genome. There are two such classes, known as **minisatellites** and **microsatellites**. Both types have proved extremely important in the construction of maps of the human genome, which is discussed in Chapter 3. Minisatellites were also the original basis of **genetic fingerprinting** used in forensic science.

Minisatellites consist of sequences between 10 and 100 bp long repeated in tandem arrays that vary in size from 0.5 to 40 kb. They tend to occur near telomeres although they have been found elsewhere. Some, but not all, minisatellites show variation in repeat number so they are sometimes referred to as **variable number tandem repeats** (VNTRs). An individual who carries two different alleles that vary in size is said to be **heterozygous**. A locus that commonly exists in different forms is said to be **polymorphic**. As we shall see in Chapter 3, polymorphic loci such as minisatellites may be used as markers to construct genetic maps, although their tendency to be clustered near telomeres limits their use as a genomic mapping marker. Some minisatellite loci are hypervariable, for example alleles of a locus called D1S8 contain between 120 and 1000 repeats of a 29 bp sequence, resulting in a variation in size from 3.3 to 29 kb. As well as variation in repeat number, the sequence of the repeat unit itself can vary in different members of the repeat array, so that loci that are **monomorphic** for length may still be highly polymorphic in structure (see Figure 9.7). The variability of minisatellite loci forms the basis of genetic fingerprinting used in forensic science and paternity testing (see Chapter 10).

Microsatellites consist of tandem repeats of units two to four nucleotides in length. They are also known as **short tandem repeats** (STRs). They are found at all locations within the genome, even within protein-coding sequences. Like minisatellites, the number of repeats in each microsatellite varies, changing its overall size. Thus microsatellites are polymorphic and they have proved to be valuable genetic markers. They have been more useful in this respect than minisatellites because their genomic distribution is more uniform. The most commonly used microsatellite for this purpose has the structure $(CA)_n$. Tetrameric STRs form the basis of the current method of DNA profiling for forensic casework (see Chapter 10).

2.2.4 Telomeres

At the end of each chromatid is a structure known as the **telomere.** The telomere consists of a large and variable number of tandem repeats of the sequence TTAGGG and has a protruding 3' end. This 3' end does not have the usual properties of single-stranded DNA. One possibility is that the free 3' end folds back on itself to form a hairpin structure through unusual base-pairing G residues (Figure 2.7). Other more elaborate structures have been proposed. The telomere is synthesised by an RNA-containing enzyme called **telomerase**. The RNA molecule in telomerase is used as a template to extend the free 3-OH end of the chromosome.

The telomere serves two essential functions.

1. Free ends of chromosomes produced by breakage are highly unstable and fuse with a high frequency to other chromosome ends. The telomere stops this happening to the ends of normal chromosomes by binding specific proteins to form a protective cap.
2. Telomeres solve a problem of replicating linear DNA molecules. DNA synthesis proceeds in the 5' to 3' direction by extending an RNA primer. DNA synthesis that uses as a template the DNA strand with the free 3-OH group at the end of a chromosome requires a primer upstream of the first base to be copied, which of course can not exist. This problem is solved by extending the 3' end using telomerase.

The telomere may play an important part in both the ageing process and cancer. Human cells are **mortal**: they can only go through about 50 divisions before they stop dividing and die. As a cell ages its telomere becomes shorter. It is possible that cell mortality occurs when the telomere becomes too short to function efficiently and vital sequences in the chromosome are damaged. One characteristic of cancer cells is that they are **immortal** and their telomeres do not shorten. It has been claimed that some cancer cells

Figure 2.7 Possible hairpin loop structure of telomeres. Unusual base pairing allows a hairpin structure to form protecting the ends of the DNA molecule.

have a much higher level of telomerase and that this is responsible for their immortality. This field is highly controversial, partly because of the formidable technical problems in the biochemical assay of telomerase.

2.3 Structure of chromosomes

The 3000 Mb of DNA that constitutes the human genome is divided between 23 pairs of chromosomes; 22 of these chromosomes are found in both males and females and are known as **autosomes**. One pair of chromosomes is different in each sex and these are known as the sex chromosomes. Females have two copies of the **X chromosome**, while males have one X chromosome and one **Y chromosome**.

2.3.1 Human karyotype

During the prophase of meiosis and mitosis, chromosomes become more condensed than in interphase. At mitotic metaphase the chromosomes are visible in the light microscope and may be examined after they have been spread on a slide. The number, size and shape of the chromosomes is referred to as the **karyotype** (Figure 2.8).

Further detail may be revealed by **banding**. In this process the metaphase spread is subject to light digestion with an enzyme such as trypsin, which breaks down proteins. It is then treated with **Giemsa**, a dye that binds DNA. Giemsa binds with different intensities at different points along the chromosome producing a series of bands called **G-bands**. The functional significance of these, if any, is not known. They may be a reflection of long-term variation in the proportion of GC base pairs along the length of the chromosome. The banding pattern is reproducible and is characteristic of each chromosome. Thus each chromosome, and even each part of a chromosome, may be recognised by its banding pattern (Figure 2.8). This provides the basis of a mapping procedure and allows changes to chromosomes to be recognised (Figure 2.9). Treatment of the metaphase spread with the fluorescent dye **quinacrine** produces a different type of banding pattern known as **Q-bands**. A karyotype of a human male treated to produce G-bands is shown in Figure 2.8.

At metaphase the chromosomes are **bivalent**. The two **chromatids** are joined at a constriction called the **centromere**. A protein structure called the **kinetochore** is located at the centromere and is the point at which the chromosome is attached to the mitotic spindle. The centromere divides the chromosome into two arms, the larger of which is called the **q arm** and the smaller the **p arm**. If the centromere is centrally placed the chromosome is said to be **metacentric**; where it is nearer one end the chromosome is said to be **submetacentric**. Where the centromere is near one end the chromosome is said to be **acrocentric**; when it is at one end it is said to be **telocentric**.

The chromosome number, the chromosome arm and the G-bands are used together to specify the location of genetic elements. Figure 2.9 shows a

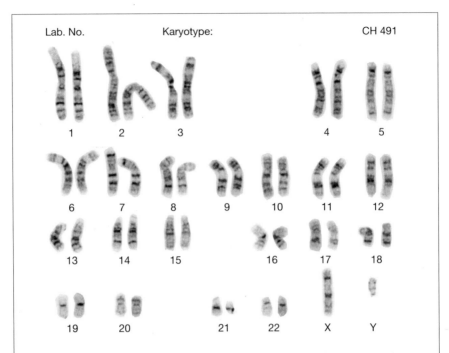

Lab. No.   Karyotype: CH 491

1   2   3   4   5

6   7   8   9   10   11   12

13   14   15   16   17   18

19   20   21   22   X   Y

Figure 2.8 Karyotype of a human male. In the metaphase spread the different chromosomes are randomly arranged. The metaphase spread may be photographed, the individual chromosomes cut out and arranged so that each chromosome is paired with its homologue. It can be seen that chromosomes vary in size. By convention the largest pair are called chromosome 1 and the smallest chromosome 22, although it has subsequently been shown that chromosome 21 is in fact smaller than chromosome 22.

map of the X chromosome by way of illustration, for example the DMD locus is said to be located at Xp21.

Fluorescence *in situ* hybridisation

Chromosomes attached to a microscope slide in a metaphase spread can be hybridised with complementary sequences in probes, allowing the chromosomal origin of the probe to be identified. This process is known as *in situ* **hybridisation**. Originally it was carried out using ^{3}H-labelled DNA. This was technically demanding and needed the use of autoradiograms that needed long development times. The procedure was revolutionised by the use of DNA probes that were labelled using different coloured **fluorophores**. These were visualised using a fluorescence microscope that excited fluorescence using the particular excitation wavelength of the fluorophore. The use of fluorescently labelled DNA probes in this way is called **fluorescence *in situ* hybridisation** (FISH). By using probes labelled with different coloured fluorophores it is possible to visualise multiple regions. FISH is an important tool in physically mapping the human genome because it allows the chromosomal origin of any cloned piece of DNA to be determined. Plate 1 shows an example of this where a probe for the elastin

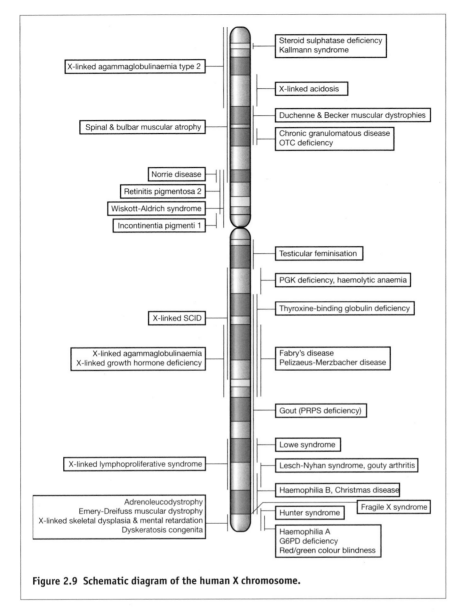

Figure 2.9 Schematic diagram of the human X chromosome.

locus hybridises to chromosome 7. The resolution of this technique is several megabases because chromosomes are in a highly condensed state in metaphase. Recently, hybridisation to interphase chromosomes or to stretched chromosome fibres has allowed high-resolution mapping where the resolution is 50 kb or even higher.

FISH has also proved to be a highly effective tool for karyotyping. Chromosome-specific probes can be used as 'chromosome paints' that allow the rapid identification of chromosomes without the careful characterisation of G-banded karyotypes. Plate 2 shows how this can be used for sex determination of a foetus using whole, uncultured cells from amniotic fluid. This is a powerful technique where there is a risk of a sex-linked disease. Plate 3 shows how chromosomes in a metaphase spread may be visualised using a chromosome paint. Chromosome paints are used extensively in the analysis of cancer cell karyotypes, which can show characteristic rearrangements during the course of the disease.

2.3.2 Packaging DNA into chromosomes

A single DNA molecule runs the length of each chromosome. This molecule is about 250 Mb in size in the largest of the chromosomes (chromosome 1). Physically the DNA in chromosome 1 is about 8 cm long, yet it is packaged at mitotic metaphase into a chromosome only 8 µm in length, a compaction ratio of about 10 000. How is this done?

The DNA is wound around itself in a hierarchy of coils and supercoils. At the deepest level, the DNA is wound around an octamer composed of two molecules each of the four different basic proteins called **histones** H2A, H2B, H3 and H4 (Figure 2.10). This basic unit is called a **nucleosome** and in the electron microscope the DNA has the appearance of beads on a string. Each nucleosome occupies about 200 bp of DNA and is linked to the next by a short stretch of naked DNA. The register of DNA in nucleosomes may be very important in the control of gene expression.

DNA wound up in nucleosomes is coiled again to form a fibre 30 nm in diameter (Figure 2.11). The 30 nm fibre is stabilised by a fifth histone called histone H1. Finally, at mitosis the 30 nm fibre winds around a scaffold of protein to from a fibre 700 nm in diameter (Figure 2.12). Various non-histone proteins are also associated with the DNA. The whole DNA–RNA–protein complex is known as **chromatin**.

During interphase, chromosomes are not so tightly condensed as metaphase chromosomes but the DNA must still be highly packaged. The transcription machinery must gain access to the DNA. Simple cartoons that depict transcription as an RNA polymerase moving along a DNA molecule

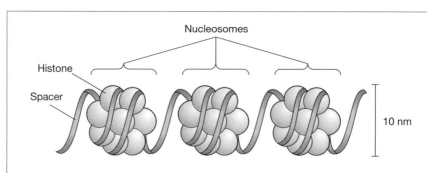

Figure 2.10 Structure of nucleosomes. DNA is wound around an octamer consisting of two copies each of four histones: H2A, H2B, H3 and H4.

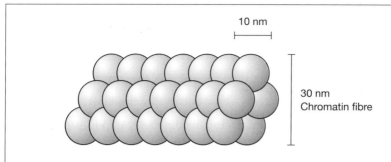

Figure 2.11 The 30 nm fibre formed by the coiling of DNA packaged into histones.

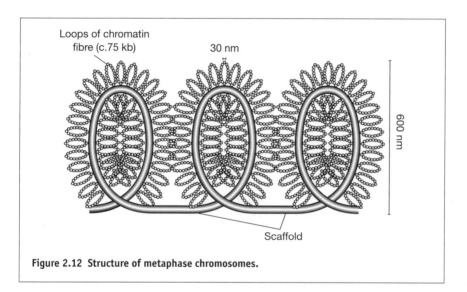

Figure 2.12 Structure of metaphase chromosomes.

like a train on a railway track are distorting reality to the point where it is no longer a valid model. It is more likely that there are about 200 RNA factories within the nucleus of the human cell. The DNA is moved through these factories rather than the other way round. Perhaps a more accurate analogy is the way a tape in an audio cassette is fed through the playing head. When the tape is not being played it must be stored in a highly packaged fashion. What happens when the tape is not properly wound inside the cassette is something we are all familiar with!

2.4 Summary

- The haploid human genome consists of 3000 million base pairs (3000 Mb) of DNA.

- It consists of different sequence classes:

- ○ Single sequence or low-copy DNA.
- ○ Intermediate repeated DNA sequences consisting of families of DNA sequences repeated between 10^2 and 10^5 times per genome.
- ○ Highly repetitive or satellite DNA, repeated 10^6 times per genome.

- Coding sequences represent only about 3% of the genome.

- Not all genes encode proteins; some encode functional RNA molecules.

- There are multiple copies of genes encoding rRNA and histones. These are arranged in long tandem repeat arrays.

- There are two different classes of intermediate repeat sequences called SINEs and LINEs. SINEs and LINEs are mobile genetic elements that may be selfish or parasitic since they undergo selection to increase in number but do not contribute to the phenotype.

- The most common LINE is called L1. It encodes a reverse transcriptase that mediates transposition to a new location by a process known as retrotransposition.

- The most common SINE is a family of sequences called *Alu* which, like the L1 family, has also spread through the genome by a process of retrotransposition.

- Most highly repetitive DNA is found in long tandem repeats at centromeres. The sequences bind proteins of the kinetechore where the spindles are attached to chromosomes at mitosis and meiosis.

- Minisatellites consist of tandem repeats of units 10–100 bp in size. The number of repeats and the sequence of the repeat units themselves are both polymorphic, making them useful genetic markers. They form the basis of genetic fingerprinting.

- Microsatellites are interspersed throughout the genome. They consist of tandem repeats of units 2–4 bp in size. They are also widely used as genetic markers.

- The number, size and appearance of chromosomes is called the karyotype. There are 22 autosomes and one pair of sex chromosomes.

- G-bands allow chromosomes and chromosome arms to be recognised.

- The telomere is a special structure that protects the ends of chromosomes. It may be important in cancer and ageing.

- The DNA molecules in chromosomes are physically long and so have to be tightly packaged.

- The DNA is wound around histones to form nucleosomes

- The nucleosomes are coiled to make a fibre 30 nm in diameter.

- At mitosis and meiosis the 30 nm fibre is coiled around a scaffold of proteins to produce a 700 nm fibre.

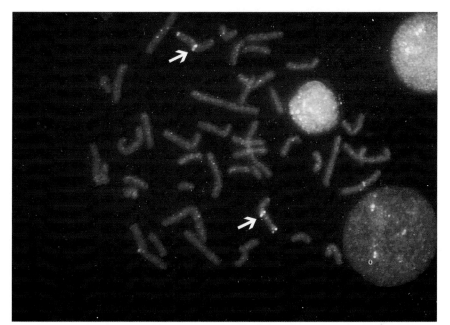

Plate 1 A gene probe for the elastin locus on the long arm of chromosome 7 is used to identify a deletion of the locus in a patient with Williams syndrome. (A) The normal karyotype. The two copies of chromosome 7 are arrowed. The signal near the telomere is a chromosome-specific probe to identify the chromosome; the signal near the centromere is the elastin locus. (B) One of the two copies of chromosome 7 (arrowed) lacks the signal near the centromere, showing the presence of a deletion. Williams syndrome affects about 1 in 10 000 of the population. It is characterised by an elfin-like appearance, hyperactive 'cocktail party' personality and multisystem disorders including multiple cardiac abnormalities, childhood hyperglycaemia, growth retardation, learning difficulties and neuromuscular disorders. It is caused by hemizygosity of the elastin locus. Elastin is a connective tissue protein that is a major component of elastic fibres in blood vessels, ligaments, etc.

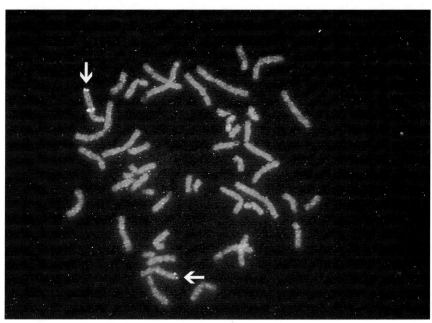

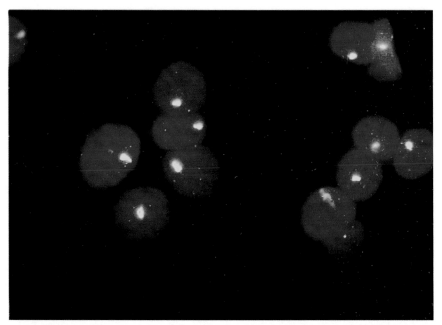

Plate 2 Probes specific for X and Y chromosones can be used to identify the sex of cells with interphase nuclei. This example shows a male foetus identified by the use of a Y-specific probe on uncultured cells from amniotic fluid.

Plate 3 A metaphase spread 'painted' with a whole-chromosome paint for chromosome 2.

Further reading

General

GRIFFITHS, A.J.F., MILLER, J.H., SUZUKI, D.T., LEWONTIN, R.C. and GELBART, W.H. (1996) *An Introduction to Genetic Analysis*, 6th edn. W.H. Freeman, New York.

LEWIN, B. (1997) *Genes VI*. Oxford University Press, Oxford.

LODISH, H., BALTIMORE, D. and DARNELL, J.E. (1995) *Molecular Cell Biology*, 3rd edn. W.H. Freeman, New York.

SNUSTAD, D.P., SIMMONS, M.J. and JENKINS, J.B. (1997) *Principles of Genetics*. John Wiley and Sons, New York.

Reassociation experiments

BRITTEN, R.J. and KOHNE, D.E. (1968) Repeated sequences in DNA. *Science*, **161**, 529–540.

Selfish DNA

DOOLITTLE, W.F. and SAPIENZA, C. (1980) Selfish genes, the phenotype paradigm and genome evolution. *Nature*, **284**, 601–603.

ORGEL, L.E. and CRICK, H.C. (1980) Selfish DNA: the ultimate parasite. *Nature*, **284**, 604–607

Transcription

LOO, S. and RINE, J. (1995) Silencing and heritable domains of gene expression. *Annual Review of Cell Biology*, **11**, 519–548.

MANIATIS, T., GOODBOURN, S. and FISCHER, J.A. (1987) Regulation of inducible and tissue-specific gene expression. *Science*, **236**, 1237–1245.

MITCHELL, P. and TIJAN, R. (1989) Transcriptional regulation in mammalian cells by sequence specific DNA binding proteins. *Science*, **245**, 371–378.

PABO, C.T. and SAUER, R.T. (1992) Transcription factors: structural families and principles of DNA recognition. *Annual Review of Biochemistry*, **61**, 1053–1095.

ZAWEL, L. and REINBERG, D. (1993) Initiation of transcription by RNA polymerase II: a multi-step process. *Progress in Nucleic Acid Research and Molecular Biology*, **44**, 67–108.

Splicing
Kramer, A. (1996) The structure and function of proteins involved in mammalian pre-mRNA splicing. *Annual Review of Biochemistry*, **65**, 367–410.

PADGETT, R.A., GRABOWSKI, P.J., KONARSKA, M.M. *et al.* (1986) Splicing of messenger RNA precursors. *Annual Review of Biochemistry*, **55**, 1119–1150.

α Satellite sequences, centromeres and human artificial chromosomes
HARRINGTON, J.J., VAN BOKKELEN, G. MAYS, R.W., GUSTASHAW, K. and WILLARD, H.F. (1997) Formation of *de novo* centromeres and construction of first generation human artificial microchromosomes. *Nature Genetics*, **15**, 345–355.

SCHULMAN, I. and BLOOM, K.S. (1991) Centromeres: an integrated protein/DNA complex required for chromosome movement. *Annual Review of Cell Biology*, **7**, 311–336.

WILLARD, H.F. (1991) Evolution of alpha satellite. *Current Opinion in Genetics and Development*, **1**, 509–514.

LINEs
HATTORI, M., KUHARA, S., TAKENAKA, O. and SAKAKI, Y. (1986) L1 family of repetitive elements in primates may be derived from a sequence encoding a reverse-transcriptase related protein. *Nature*, **321**, 625–627.

Retrotransposons
ROGERS, J. (1986) The origin of retroposons. *Nature*, **319**, 725.

WAGNER, M.A. (1986) Consideration of the origin of processed pseudogenes. *Trends in Genetics*, **2**, 134–136.

***Alu* elements**
BRITTEN, R.J. (1996) Evolution of *Alu* retroposons. In *Human Genome Evolution*, Jackson, M., Strachan, T. and Dover G. (eds), pp. 211–228. Bios Scientific Publishers, Oxford.

OKADA, N. (1991) SINEs. *Current Opinion in Genetics and Development*, **1**, 498–504.

ULLU, E. and TSCHUDI, C. (1984) *Alu* sequences are processed 7SL RNA genes. *Nature*, **312**, 171–172.

Minisatellites

ARMOUR, J.A.L. (1996) Tandemly repeated minisatellites: generating human genetic diversity via recombinational mechanisms. In *Human Genome Evolution*, Jackson, M., Strachan, T. and Dover G. (eds), pp. 171–190. Bios Scientific Publishers, Oxford.

JEFFREYS, A.J. (1987) Highly variable minisatellites and DNA fingerprints. *Transactions of the Biochemical Society*, **15**, 309–316.

JEFFREYS A.J. WILSON, V. and THEIN, S.L. (1985) Hypervariable minisatellite regions in human DNA. *Nature*, **314**, 67-73.

Microsatellites

HANCOCK, J.M. (1996) Microsatellites and other simple sequences in the evolution of the human genome. In *Human Genome Evolution*, Jackson, M., Strachan, T. and Dover G. (eds), pp. 191–210. Bios Scientific Publishers, Oxford.

WEBER, J.L. and MAY, P.E. (1989) Abundant class of human polymorphisms that can be typed using the polymerase chain reaction. *American Journal of Human Genetics*, **44**, 388–396.

Telomeres

ZAKIAN, V. (1995) Telomeres: beginning to understand the end. *Science*, **270**, 1601–1607

Chromatin structure

FELSENFELD, G. (1992) Chromatin as an essential part of the transcriptional mechanism. *Nature*, **355**, 219–224.

FELSENFELD, G. and McGHEE, J.D. (1986) Structure of the 30nm fibre. *Cell*, **44**, 375-377.

TRAVERS, A.A. and KLUG, A. (1987) The bending of DNA in nucleosomes and its wider implications. *Philosophical Transactions of the Royal Society of London Series B*, **317**, 537–561.

CHAPTER THREE

Mapping the human genome

Key topics

- Importance of genomic maps
- Sequence tagged site
- Genetic maps
  - Key concepts, LOD scores, informative meioses, phase, haplotype
  - RFLP, minisatellite and microsatellite mapping markers
  - The Genthon map
- Physical maps
  - Clone maps
  - Radiation hybrid map
  - STS content maps
  - FISH
  - Long-range restriction maps
- Expression maps
- Integration of genetic, physical and expression maps

3.1 Introduction

The construction of detailed maps of the human genome has arguably been the single most important development in human genetics over the last few years. The reason is simple: maps allow us to find and isolate genes when we have no other information about them apart from their location. In the last few years genes responsible for many of the major monogenic disorders have been isolated by such **positional cloning**. Furthermore, the focus of research in human genetics has now shifted to the analysis of complex diseases. Detailed maps provide the only means by which the genes that contribute to disease susceptibility can be identified and ultimately characterised.

Maps of the human genome can take different forms. The difference between them is important. Genetic maps are based on recombination frequencies between genetic markers at meiosis. Genetic markers are features of the human genome that are variable in different individuals. They may control phenotypic characteristics, which include inherited diseases, or they may be a variable feature of the genotype observed but have no phenotypic consequences. Genetic markers freely recombine at meiosis unless they are located in the same region of the genome, in which case they are said to be linked. Linked markers tend to be co-inherited; the closer they are, the higher the frequency of co-inheritance. Once a detailed genetic map has been constructed using a reference set of neutral markers, a gene responsible for a genetic disorder can be mapped by observing co-inheritance between the disease and a reference marker. Here we see the critical importance of genetic maps: they provide the only connection between biological reality and the underlying genome. In many cases, without genetic maps nothing else follows.

The location of a gene on the genetic map is only important in so far as it provides a means of identifying and isolating the DNA sequence that forms the gene. This allows us to study the nature of the protein encoded by the gene and characterise the mutations that may affect its function. Physical maps describe the location of DNA sequences. They are constructed by plotting the location of physical mapping markers based on DNA sequences. If we can align the genetic and physical maps then the genetic location of a gene directs us towards the DNA sequence that forms the gene.

Only a small fraction of DNA actually encodes proteins so often we must examine a large amount of non-coding DNA to find a gene that we are looking for. Here another form of map comes to our aid. Genes are DNA sequences that are expressed to form mRNA molecules. If we plot the origin of all mRNA molecules, then we will form a map of the DNA sequences that are expressed. Once genetic and physical maps have identified the genomic region in which a gene resides, we can restrict our attention to those DNA sequences that are expressed.

Thus genetic, physical and expression maps can be used together to find genes. Figure 3.1 illustrates the general process by which we can move from studying the segregation of a genetic disorder in a family to identifying the DNA sequences that have been altered by mutation. All of this depends on being able to align the three different map formats. Initially the different types of map were based on landmarks that were not necessarily compatible. One key advance in constructing genomic maps was the development of a marker, called a **sequence tagged site** (STS) (see below), that could be used with all three types of map. Thus when its location was defined in one map type, its location was also defined in the two other types of maps.

There is a second major reason for constructing maps. Before the human genome can be sequenced, it is necessary to assemble all the DNA into a set of ordered clones. This is essential to allow sequencing to proceed systematically instead of chaotically. As discussed below, the order of the clones itself is a form of map. Clone maps have so far been constructed in vectors called **yeast artificial chromosomes** (YACs) that carry large DNA inserts (~1 Mb).

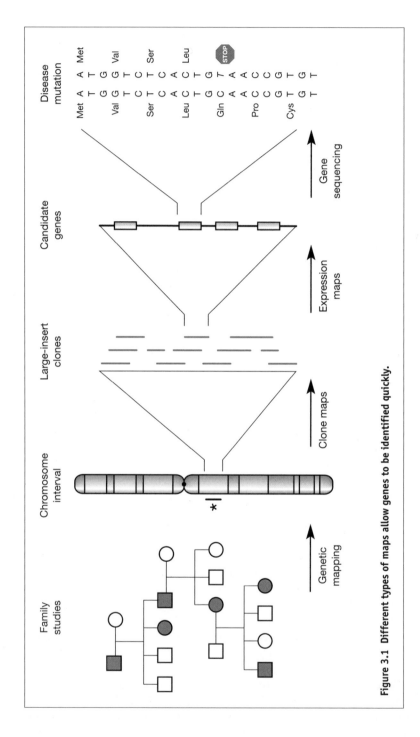

Figure 3.1 Different types of maps allow genes to be identified quickly.

This is far too large for sequencing, so it will be necessary to construct new libraries with much smaller inserts. The map of physical landmarks will allow us to quickly order these clones in preparation for sequencing. However, to do this the physical map will have to be detailed. The first goal of the Human Genome Project (discussed in Chapter 8) was to construct maps of the genome. It established targets for the resolution of these maps. These are an STS map with STS markers spaced every 100 kb and a genetic map with a resolution of 0.7 cM. As we shall see this goal for the genetic map has been achieved and the physical map is nearly complete.

3.2 Sequence tagged site: a common currency for the different types of map

Mapping markers are the landmarks whose positions are plotted to construct the different types of map. As discussed above, it is important that the different types of map can be aligned with each other. However, markers that can be used for one type of mapping cannot necessarily be used for another. For example, genetic markers must be based on features that are naturally polymorphic. A disease segregating in a family is a genetic marker because some members are affected and others are not. However, unless the disease gene has been cloned, it cannot be used as a physical marker. A physical marker could be a sequence of DNA whose presence in a clone or location on a chromosome can be determined. However, if the sequence is the same in all individuals it cannot be used as a genetic marker.

There is clearly a need for a marker that can be used in the different types of map. The **polymerase chain reaction** (PCR) is used to generate a type of physical landmark called an STS that meets this requirement. PCR amplifies DNA using synthetic oligonucleotides as primers (Box 3.1). An STS is a stretch of DNA about 300 bp in length. As its name implies, the sequence of the STS is used to tag the larger DNA molecule from which it was derived so that the DNA region can be identified wherever it occurs. Figure 3.2 illustrates how STS markers may be generated and used to order a series of clones. The nucleotide sequence of the STS is used to specify the sequence of two synthetic oligonucleotides that will bind in opposite orientations at either end of the STS. Thus they will amplify the sequence if they are used as primers in a PCR reaction. PCR using these primers can therefore be used as a test to discover if the sequence is present in a particular DNA sample.

Originally the concept of an STS marker was developed for physical mapping because it was based on recognition of a sequence. However it was subsequently realised that an STS could be polymorphic, so long as the primers were designed to anneal to constant regions flanking some form of length polymorphism. We shall see below how microsatellites can be used to provide polymorphic STSs that are used for genetic mapping. STSs can also be generated from cDNA molecules and can be used to tag sequences that are expressed (**expressed sequence tag** or EST). The STS therefore provides a type of marker that is compatible with the three major map formats and so allows them to be aligned with respect to each other.

BOX 3.1: THE PCR REACTION

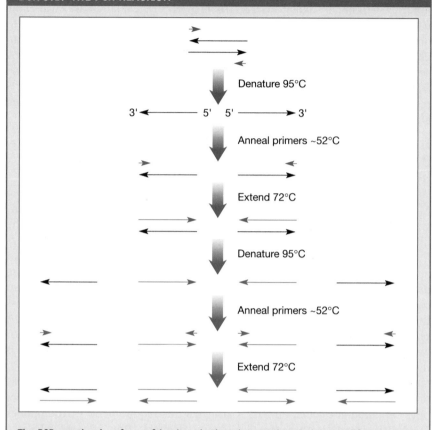

The PCR reaction is a form of *in vitro* cloning that can be used to amplify very small quantities of DNA. It depends on DNA polymerases from thermophilic bacteria, which can withstand temperatures of 95°C and have a temperature optimum of 72°C. DNA polymerases can only synthesise new DNA by extending the 3' end of a pre-existing primer. This is exploited to target new synthesis to a region between sites where two primers bind. In each cycle the template DNA is denatured by heating to 95°C, annealed to the primer at a temperature of about 52°C (see below) and then incubated at 72°C for the polymerase to extend the primer using the target DNA as a template. In the figure the target DNA is shown in black and newly synthesised DNA is shown in blue; the arrows on each DNA strand point toward the 3' end. Two complete cycles are represented, and it can be seen that in each the amount of target DNA is doubled. A total of 30 cycles means that amounts of DNA as small as a single molecule can be amplified to produce microgram quantities of DNA. The reaction is carried out in a thermocycler that can be programmed to carry out the cycles as specified by the user. The primers can be readily synthesised using automated technology. The length of the primers is usually about 20 bases; the exact length and its base composition determines the optimum annealing temperature. The annealing temperature and ionic composition of the buffer are adjusted to ensure maximum stringency.

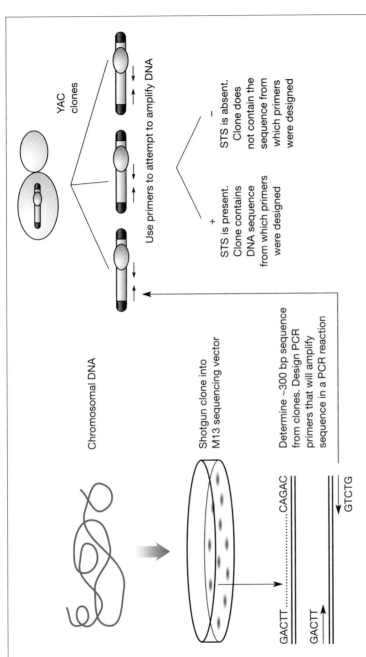

Figure 3.2 STS markers. Chromosomal DNA is randomly cloned (shotgun cloning) into an M13 vector, which allows 300 bp of sequence to be easily determined. The sequence is then used to design PCR primers that will amplify that sequence wherever it occurs. Initially these sequences are anonymous, i.e. their location and any function encoded by the sequence is unknown. However, we can use the primers to test whether the sequence is present in other types of clone. In this example YAC clones are screened using these primers (as described below, YACs are a special cloning vector that take large inserts). A positive result indicates that the clone contains the STS. If two clones contain the same STS then the inserts they carry are likely to overlap. In this way we can order the clones in a random library. This process is described in more detail later in the chapter.

Apart from the fact that they provide a common marker for different types of map, there are four other important advantages of the STS methodology.

1. They allow different sources of DNA fragments to be examined for the presence of a common sequence; for example two clones may be compared to detect any overlap between them. The sources of DNA may be totally different: we discuss below how YAC clones and radiation hybrid clones are forms of physical mapping that both use STS markers. Because of this the two types of map can be aligned with each other.
2. If a researcher in one laboratory wishes to examine whether an STS is present in clones they are working on, they do not have to physically obtain a clone from another laboratory that produced the STS. Instead the sequence of the primers is used to synthesise the oligo-nucleotides using a widely available technology for automatic synthesis. There is no need to physically maintain and distribute stocks of DNA. This solves an important logistical problem of maintaining and distributing clone libraries.
3. The PCR procedure requires both primers to bind at the correct distance apart *and* in the correct orientation. This reduces the number of false positives that can arise using other methods to compare DNA molecules, such as hybridisation.
4. The PCR test lends itself to automation, allowing very large numbers of tests to be performed. Since the human genome is so large this is an important practical consideration.

3.3 Genetic maps

Genetic maps are based on the order of genetic mapping markers and the genetic distance between them measured by the recombination frequency. There are a number of concepts connected with genetic maps that we need to consider before we discuss the construction of a comprehensive map of the human genome.

Map units
All maps must have a unit of scale. The unit of human genetic maps is the **centimorgan** (cM). 1 cM is defined as a recombination fraction of 0.01, i.e. 1 in 100 gametes will be recombinant and the remaining 99 will have the parental configuration. Physically, 1 cM corresponds to between 0.7 and 1 Mb of DNA sequence, but there is no invariant relationship between physical and genetic distances (see Figure 3.9).

Informative and non-informative meioses
In order to measure the recombination frequency between two markers at meiosis it is necessary that the meiosis is **informative**, i.e. it is possible to distinguish between parental and recombinant chromosomes. This requires that both markers are heterozygous. Figure 3.3 illustrates why this is the case.

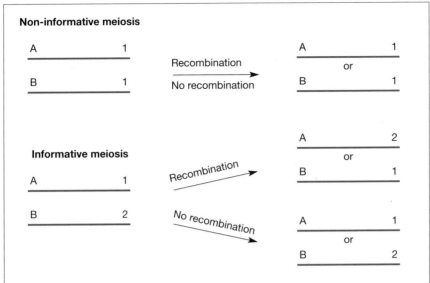

Figure 3.3 Informative and non-informative meioses. *Left*: two homologous chromosomes in a diploid cell about to undergo meiosis. Two adjacent loci, each with two alleles, are located on these chromosomes. *Right*: the structure of haploid gametes that result from the meiosis is shown. *Top*: the meiosis is uninformative because one of the loci is homozygous; after meiosis the gametes will have chromosomes that are identical whether or not a recombination event has occurred. *Bottom*: recombination results in chromosomes that can be distinguished from non-recombinant chromosomes; the meiosis is therefore said to be informative.

Phase
Phase specifies whether particular alleles at adjacent loci are on the same (*cis*) or different chromosomes (***trans***). For example, in the informative meiosis depicted in Figure 3.3, alleles A and 1 are in *cis* while alleles A and 2 are in *trans*. Clearly it is necessary to know the phase in order to know which outcomes of the meiosis represent recombinant chromosomes and which represent parental chromosomes. Figure 3.4 shows an example of how the phase can be established by examining three or more generations.

Haplotype
A **haplotype** is a set of closely linked alleles that tend to be inherited together at meiosis, i.e. not separated by recombination ('haplotype' is a contraction of 'haploid genotype'). In the previous example there are two parental haplotypes, A1 and B2. Haplotypes may consist of many more than two alleles. Generally, alleles making up a haplotype will be inherited as a block because their close proximity makes it unlikely they will be separated by recombination.

CEPH families
Recent genetic maps have been constructed using a set of reference families. These consist of samples collected from Mormon families living in Utah, USA

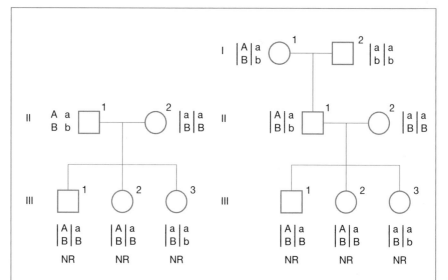

Figure 3.4 Phase can be worked out in a three-generation family. The figure shows the segregation of two linked loci each with two alleles (A/a and B/b). In the two-generation family on the left, the phase of the alleles in individual II-1 is unknown; therefore it cannot be established whether the progeny in generation III have inherited recombinant or parental chromosomes from II-1. The genotype of the grandparents allows the phase of the alleles in II-2 to be established. This reveals whether individuals in generation III have inherited parental or recombinant chromosomes. The vertical lines indicate the phase of the alleles where it is known.

and French–Venezuelan families. Mormon families are typically much larger than normal and the keeping of detailed family records is part of the Mormon culture. The samples were used to establish permanent cell lines, which are maintained by the Centre d'Étude Polymorphism Humain (CEPH) in Paris. Each family consists of three generations with four grandparents, two parents and at least six children. Three generations allows the phase of markers in the parents to be established. The segregation of markers in the six children allows the results of meiosis in the parents to be determined. These cell lines provide an invaluable source of reference pedigrees, which are used in many different sorts of investigation into the human genome.

LOD score analysis
In experimental organisms, a large number of progeny from a genetic cross may be examined to accurately measure recombination frequency. Clearly this is not possible in human families. This problem is overcome by using a special form of mathematical analysis known as log ratio of odds or **LOD score** analysis (Box 3.2). The result of LOD score analysis is a numerical value that measures the ratio of odds that two loci are linked at given recombination fraction θ, compared with the chances that they are unlinked ($\theta = 0.5$). The threshold for declaring linkage is an LOD score of

BOX 3.2: LOD SCORE ANALYSIS

An LOD score is defined as the logarithm of the ratio of odds (Z) of the observed outcome if two loci are linked with a recombination fraction θ, to the odds of the observed outcome if they are unlinked ($\theta = 0.5$). Thus

$$Z_x = \frac{\text{Odds of observed result if } \theta = x}{\text{Odds of observed result if } \theta = 0.5}$$

$$\text{LOD}_{\theta = x} = \log_{10} Z_x$$

The calculation is then repeated using different values of θ. The value of θ when LOD is at a maximum is called the MLS. The simplest application of LOD score analysis compares two markers, but it may be adapted to analyse multiple loci. This is known as multipoint LOD score analysis.

Statistical significance

Prior to the observation, the odds that any two loci are linked is about 1 in 50. This is known as the **prior probability**. The odds that they are linked based on observation is known as the **conditional probability**. This is what is measured by the LOD score. The overall odds that they are linked is the product of the prior and conditional probabilities. This is known as the **posterior probability**. Conventionally, for a result to be judged statistically significant the odds of it being true should be equal to or greater than 20:1. Therefore an LOD score of 3 is required (i.e. $Z = 1000:1$) for statistical significance because:

Posterior probability = prior probability x conditional probability
$$= 1:50 \times 1000:1$$
$$= 20:1$$

A worked example

Recall the three-generation pedigree from Figure 3.4 and consider the genotypes of the children in generation III. Are loci A and B linked? Assume that it is possible to determine the status of both alleles at both loci as shown overleaf. All the children must have inherited a chromosome with the genotype aB from parent II-2. The other chromosome must have come from parent II-1 and can be categorised as recombinant (R) or non-recombinant (NR). In fact, in this example, they are all non-recombinant.

$\theta = 0$ (the loci are completely linked)

For each child there is a 0.5 chance of the observed result, because the child inherits one of two chromosomes. For the whole family, the odds of the observed result is $0.5^3 = 0.125$.

If $\theta = 0.5$ (the loci are unlinked): for each child there is a 0.25 chance of the observed genotype, because there are four genotypes and each is equally likely. For the whole family, the odds of the observed result is $0.25^3 = 0.015625$. Thus

$$Z_{\theta = 0} = \frac{0.125}{0.015625} = 8$$

$$\text{LOD}_{\theta = 0} = 0.9$$

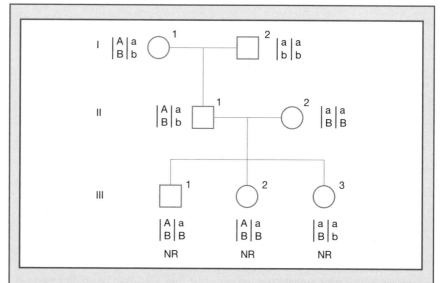

If θ = 0.25

For each child the odds of the observed result is 0.5 × 0.75 = 0.375, because the odds of inheriting a non-recombinant chromosome is 0.75 and there are two possible chromosomes that could be inherited. For the whole family, the odds of the observed result is $0.375^3 = 0.053$.

If θ = 0.5: for the whole family, the odds of the observed result = 0.015625 (as above).

$$Z_{\theta = 0.25} = \frac{0.053}{0.015625} = 3.39$$

$$LOD_{\theta = 0.25} = 0.53$$

The calculation is then repeated for other values of θ (0.1, 0.2, 0.3, etc.).

From this limited calculation we conclude that MLS = 0.9 when θ = 0, i.e. the alleles appear to be linked but the result falls short of significance.

We could now take account of results from other families. The odds of seeing an observed result in more than one family is the product of the odds in the individual families. Since LOD scores are logarithmic, LOD scores from different families can be directly added together. Thus if a similar LOD score was seen in three to four other families, the combined LOD score would become significant (four similar families would give an LOD score of 3.6).

This example is highly idealised. For example, the phase and genotypes are known. In practice, more complicated mathematical models are used and it is possible to calculate LOD scores for imperfect situations, for example if the phase is not known or penetrance of a disease is incomplete. The penalty is that the resulting LOD scores are lower so that more data need to be collected for significance. In the above example, if the phase is not known (i.e. only the two-generation family in Figure 3.4 was available) then MLS = 0.62 when θ = 0.

3 or more. It is unlikely that an LOD score of 3 will be obtained from one
family. However, the mathematical properties of the test allow data from a
number of families to be combined.

A related parameter to the LOD score is **maximum likelihood score**
(MLS). This is the LOD score for the most likely of a series of alternatives.
For example, an MLS score can be calculated for the most likely recombination fraction (θ) between two markers. This is done by iterating the LOD
score calculation assuming a different value of θ each time. The MLS
obtained gives the most likely value of θ. MLS can also be calculated when
other parameters of a genetic model are varied, i.e. degree of penetrance
(see Chapter 5). In practice, LOD scores and MLS values are derived by
computer analysis of the observed results. One popular program for this
purpose is called LIPED.

3.3.1 Mapping markers

The first maps of the human genome were produced in the 1970s and were
based on linkage between markers such as enzyme polymorphisms, disease
genes, blood groups and other phenotypic traits that were monogenic in
origin. Linkage between such markers was rare so the resolution of the
maps was low and large areas of the genome were not represented. Progress
in the construction of human genetic maps was blocked because of a lack
of usable markers in relation to the size of the genome

Restriction fragment length polymorphism

The breakthrough came in 1980 with the realisation that many sequences in
the genome were polymorphic in a way that did not have phenotypic effects.
Such markers are called **neutral molecular polymorphisms**. The first such
marker to be used is called a **restriction fragment length polymorphism**
(RFLP). It is based on the presence or absence of a target for a restriction
enzyme, usually due to a polymorphism at a single base pair (Figure 3.5).

RFLPs allowed the construction of a comprehensive map for the first
time in 1987. This map had an enormous impact on human genetic research
and was the basis for the cloning of many of the major monogenic disease
loci. RFLP markers also provided an important tool for prenatal diagnosis,
allowing a chromosome carrying a disease locus to be tagged and its segregation in a family to be followed.

Minisatellites

The problem with RFLP markers is that by definition there can be only two
alleles at a locus, i.e. the restriction site is either present or absent. The
maximum proportion of individuals in a population that are heterozygous is
only 50%. Since genetic markers are only informative when they are heterozygous, on many occasions an RFLP marker will not provide useful
information. In practice the situation is much worse, because usually one
allele is less common and the frequency of heterozygotes is less than 50%.
The proportion of meioses using a particular marker that will be informa-

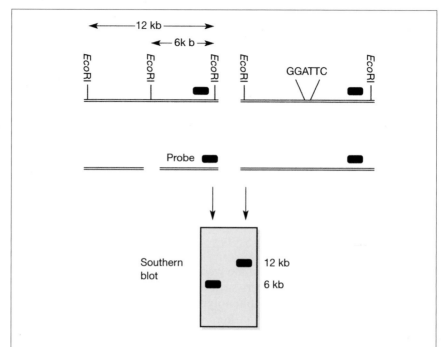

Figure 3.5 An RFLP marker. Parts of DNA molecules from two chromosomes differ from each other by a single base pair, which results in the absence of an *Eco*RI site in one of the chromosomes. Upon digestion with *Eco*RI, the chromosome without the extra *Eco*RI site produces a larger fragment than the other chromosome. The difference is recognised by Southern hybridisation using a probe (thick line) that hybridises within the region encompassed by the two flanking *Eco*RI sites present in both molecules.

tive is called its **polymorphism information content** (PIC). The maximum PIC value of a biallelic marker is 0.375 (it is less than 0.5 because some meioses are not informative even if both parents are heterozygous). Minisatellite and microsatellite markers described below typically have much higher PIC values.

A more useful type of marker is one that is naturally multiallelic, so there is a much higher chance that an individual will be heterozygous. Minisatellite loci, also known as VNTRs (see Chapter 2), meet this criteria. The number of repeats in the locus is usually determined by Southern hybridisation. Figure 3.6 illustrates how the segregation of a VNTR may be followed in a family.

Microsatellites

Minisatellites have been successfully used in the construction of genetic maps. However, their use is limited by their distribution in the genome, as they tend to be clustered near telomeres. Furthermore, the Southern blot procedure needed to score them is both laborious and expensive. PCR can be used but the large size of some minisatellites makes them difficult to amplify. A more useful type of marker is the microsatellite (see Chapter 2),

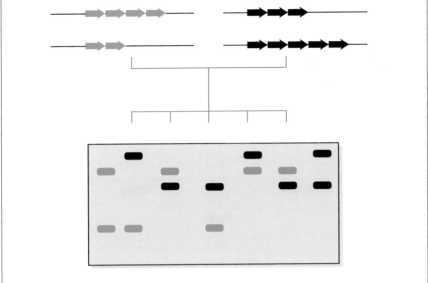

Figure 3.6 Segregation of a VNTR genetic marker in a family. The length of the VNTR depends on the number of repeats present. This can be recognised in a Southern blot using a sequence unique to the minisatellite as a probe. The genotypes of the two parents are shown in the left and right lanes respectively. Note that there are four alleles at this locus segregating in a Mendelian fashion, each child receiving one allele from each parent. This is indicated by colour coding.

which is more common and more evenly distributed than VNTRs. The microsatellite based on CA repeats has come to be the standard in the construction of genetic maps. The length of microsatellites is determined by PCR using primers designed from surrounding unique-sequence DNA (Figure 3.7). This makes them a special form of STS, which allows genetic maps based on microsatellite markers to be aligned with the different forms of physical map.

3.3.2 A comprehensive genetic map of the human genome

To be useful, a genetic map must be both comprehensive and detailed. Comprehensive means that it extends over the entire genome with no regions left unmapped. Detailed means that the distance between the markers is small, so that it is easy to detect linkage between the markers and biologically relevant loci, such as those which contribute to a familial disorder. Microsatellite loci have been used to construct a comprehensive map of the human genome, the resolution of which meets the goal set by the Human Genome Project. This map, constructed by the **Genethon** laboratory in Paris, is regarded as the definitive genetic map of the human genome.

The Genethon map is based entirely on microsatellite markers. The process of constructing the map began with identifying and sequencing 5 264 CA repeat loci. The genotype of these loci was then determined in

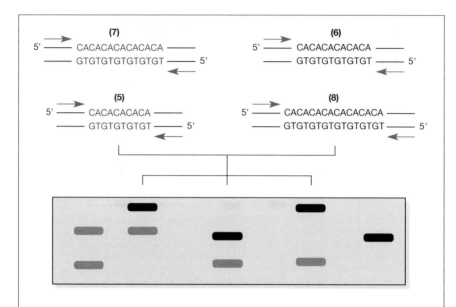

Figure 3.7 Segregation of a microsatellite locus in a family. The microsatellite locus is amplified using the PCR reaction with primers that anneal to the unique sequence either side of the CA repeats (shown by arrows). The size of the fragment produced depends on the number of CA repeats at each locus, indicated by the figure in parentheses above each DNA molecule. The products of the PCR reaction are separated by polyacrylamide gel electrophoresis, which is capable of resolving the difference in size between the different alleles. Note that there are four different alleles at the locus which show Mendelian segregation, each child receiving one allele from each parent (shown by colour coding).

members of large CEPH families. To map the autosomes a total of 134 individuals in eight families were examined, allowing the outcomes of 186 meioses to be analysed. The mapping of the X chromosome required the examination of 20 families, 304 individuals and 291 meioses.

Once the genotypes of the markers had been obtained they were analysed first to determine the order of the markers and then the genetic distance between them. This is done using specially developed computer programs. The 5 264 markers analysed were found to define 2 335 positions. The order of 2 032 of these was determined with odds of 1000:1 against alternative orders. One of the most difficult and time-consuming steps in this process was checking for errors.

The length of the map in females is 4 396 cM but only 2 769 cM in males. The difference is due to an elevated rate of recombination in female meioses compared with male meioses. The sex-averaged length is 3 699 cM. The length of the genetic maps for individual chromosomes generally reflects their physical size. The map of the largest, chromosome 1, is 292 cM; the smallest, chromosome 21, is 58 cM.

The average distance between markers is 1.6 cM. An important consideration is how evenly spaced these markers are. There are still some parts of the genome where there are not many markers, although these are a small minority of the total. For example, in 1% of the genome the distance between markers is over 10 cM. This actually consists of three positions where the interval between adjacent markers is 11 cM. This could be due to a large physical distance separating the markers, although it could also be due to enhanced rates of recombination in a particular region of the genome. In fact the physical map showed that at least part of the distance is due to increased recombination. In contrast, there were some places where markers showed no recombination with each other but were physically separated by several megabases of DNA.

Such observations emphasise that measurements of genetic map distances assume that the rate of recombination between two markers depends only on the physical distance that separates them. This requires that the frequency of recombination is constant throughout the genome. In practice this is not true. There are recombination hotspots where recombination is more frequent and coldspots where it is less frequent.

The completion of a detailed and comprehensive genetic map is a milestone in human genetics. It provides an essential resource that will be used for the rapid identification of monogenic disease loci, allows the mapping of genes that make a contribution to multifactorial disease and provides a framework to check the authenticity of physical maps. Panels of markers evenly spaced through the genome may now be purchased from biotechnology companies and these are routinely used to hunt for genes of interest.

A sample of a similar genetic map constructed by the Cooperative Human Linkage Centre is shown in Figure 3.8, which shows an overall representation of chromosome 18 on a Web page. Greater detail for any part of the map is obtained by selecting a particular chromosome interval.

3.4 Physical maps

Physical maps plot the actual location of DNA sequences in the genome. There are a number of different ways in which this can be achieved.

- **Clone maps** consist of libraries of overlapping clones where the relationship of each clone to the other clones in the library is defined.

- **Radiation hybrid (RH) maps** are based on the frequency with which markers are found on the same fragment of DNA after the genome has been fragmented by irradiation with X-rays.

- **Expressed sequence maps** plot the sites on the genome called ESTs, which are expressed in mRNA. Essentially this type of map is based on the physical location of genes. Remember that these constitute less than 5% of the total genome.

- **STS content maps** plot the location of a physical mapping marker (an STS).

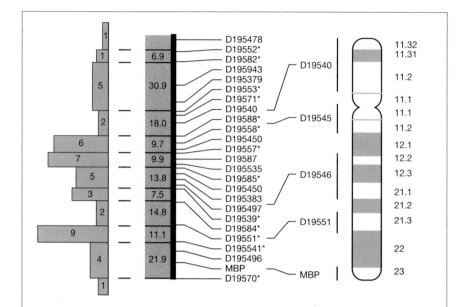

Figure 3.8 Genetic map of chromosome 18. The vertical histogram on the left shows the number of markers in each chromosome interval, shown in the centre (the figures show the size of the interval). The physical location of selected markers revealed by FISH is indicated on the cytogenetic map on the right. The map can be visited at the Web site of the Cooperative Human Linkage Centre (http://www.chlc.org/HomePage.html). Clicking one of the intervals reveals more detail of the region.

- **Long-range restriction maps** are based on the position of the target sites of rare cutting enzymes determined by pulsed field gel electrophoresis.

- FISH locates the chromosomal position of cloned fragments of DNA.

In practice these overlap because clone maps, RH maps and EST maps all make use of STSs in their construction. Moreover, as we have seen the latest form of genetic map is based on polymorphic STSs. This means that we can align all of these maps, so that it is possible to move from one format to another. This is illustrated in Figure 3.9, which should be referred to throughout the following sections. Originally the term 'physical map' was most often used to describe clone maps, but is now more often used to describe a map based on the order of STS markers. This reflects the central importance of STSs in mapping. They are used to construct clone and RH maps and can be cross-referenced to expression maps based on ESTs and genetic maps based on microsatellites.

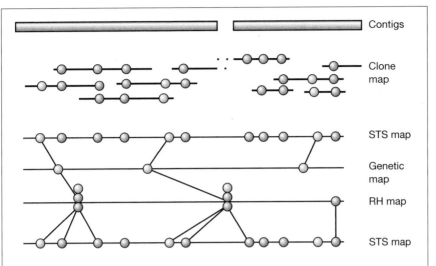

Figure 3.9 Different types of map can be aligned by STS content. Each STS is shown as a sphere; microsatellite-based STSs suitable for genetic mapping are shown in blue. The top part of the diagram shows the inserts of different clones aligned using STS content. The clones fall into two groups called contigs. Within each contig the overlap between clones may be traced through shared STSs, thus defining the spatial relationship of each clone to all the other clones in the contig. The extent of each contig is shown by the shaded bars. It is possible that the two contigs overlap (dotted line) but this has not been revealed by STS content. The overlap between the contigs allows the position of the STSs to be plotted to form the STS content map (shown twice, on the first and fourth line). Genetic and RH maps are also based on STSs, allowing these maps to be aligned with the STS content map (shown by fine lines). Note that the distortion of map intervals in the genetic map is caused by variation in recombination frequencies. RH and genetic maps can link STSs together over long distances but the order of closely linked STSs may not be resolved. The genetic and RH maps serve as a framework for the clone maps. In this example, they place the two contigs next to each other in the absence of any physical overlap being detected between them.

3.4.1 Clone maps

The original concept of a clone map was to specify the relationship of the clones in a gene library by determining how they overlapped. A group of clones whose relationship is defined in such a manner is called a **contig** (Figure 3.9). Note that the overlap between clones allows the physical relationship between two non-overlapping clones to be specified because their proximity can be traced through the overlap of intervening clones.

A perfect clone map of the human genome would consist of 24 contigs, one for each of the 22 autosomes and each of the sex chromosomes. In practice this is difficult or impossible to achieve because there are always gaps where two contigs cannot be joined. This may be because some parts of the genome are unclonable so there are no clones in the library that cover the region, or because an overlap does exist but cannot be recognised by the methods employed to detect overlaps.

This problem may be solved by using other mapping formats, such as genetic maps or RH maps, to provide an overall framework for the position of the contigs and thus detect those adjacent to each other (Figure 3.9). In order to do this, it is necessary to locate the position of STS mapping markers within the clones so that they can be anchored to the framework provided by the other mapping formats.

There are three phases in the construction of a clone map:

1. construction of the genomic library;
2. detection of overlap between the clones to produce contigs;
3. closing the gap between contigs to produce a continuous map.

Constructing a genomic library

A gene library consists of a collection of clones that together contain all the sequences present in the donor genome. Figure 3.10 illustrates the process using a YAC vector. For these to be used to construct a clone map, it is essential that two conditions are fulfilled:

1. every sequence from the donor genome is present so that the library is complete;
2. the **topology** (the geometrical arrangement) of sequences in the recombinant clones faithfully reflects that in the donor genome.

The completeness of the library is affected by two factors: the number of clones in the library and whether any sequences in the donor genome are unclonable in the host. The number of clones required is a function of the size of the insert in each clone related to the size of the total genome. Generally, the larger the insert that can be carried by the vector, the fewer clones are required for completeness. For this reason YAC vectors have been used for the first clone maps of the human genome because they can carry large inserts and thus complete libraries contain far fewer clones ($\sim 10\,000$) than are necessary using other types of vector (see below).

Sequences from the donor genome may be unclonable for two main reasons. Firstly, they may not be biologically neutral in the cloning host, for example they may encode a protein whose biological activity slows cell growth. Secondly, recombination may take place between direct repeats that results in deletion of the intervening sequences (Figure 3.11). Such recombination can make it very difficult to clone highly repetitive DNA sequences arranged in tandem repeats.

The major reason why the topology may be disrupted is that fragments of DNA from different parts of the genome have ligated together to form chimeric clones. When analysed, such clones will falsely indicate that the two sequences are adjacent to each other in the genome.

These considerations determine the choice of vector for the library construction. Routine cloning plasmids, such as the pUC series, cannot be used for this purpose because the size of the inserts they can carry are far too small for the scale of the task. Vectors based on phage λ (maximum insert size, 20 kb) and **cosmids** (maximum insert size, 40 kb) have been used for many

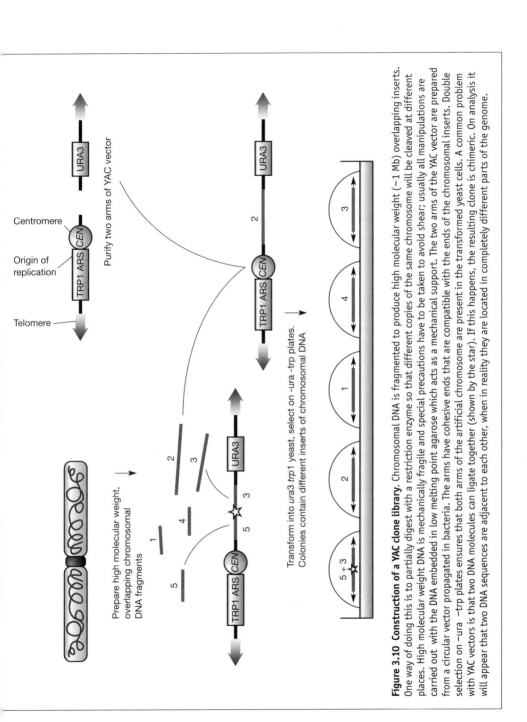

Figure 3.10 Construction of a YAC clone library. Chromosomal DNA is fragmented to produce high molecular weight (~1 Mb) overlapping inserts. One way of doing this is to partially digest with a restriction enzyme so that different copies of the same chromosome will be cleaved at different places. High molecular weight DNA is mechanically fragile and special precautions have to be taken to avoid shear; usually all manipulations are carried out with the DNA embedded in low melting point agarose which acts as a mechanical support. The two arms of the YAC vector are prepared from a circular vector propagated in bacteria. The arms have cohesive ends that are compatible with the ends of the chromosomal inserts. Double selection on –ura –trp plates ensures that both arms of the artificial chromosome are present in the transformed yeast cells. A common problem with YAC vectors is that two DNA molecules can ligate together (shown by the star). If this happens, the resulting clone is chimeric. On analysis it will appear that two DNA sequences are adjacent to each other, when in reality they are located in completely different parts of the genome.

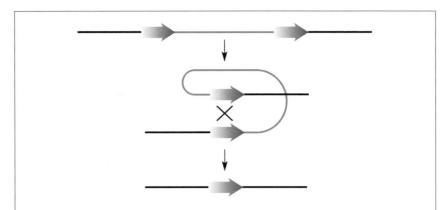

Figure 3.11 Recombination between direct repeats leads to deletions of DNA between them. Two direct repeats are shown by the arrows. DNA can loop back so that two sequences are aligned, allowing recombination to take place. As a result the intervening DNA (shown in blue) and one copy of the direct repeat is lost.

years to construct libraries. However, the maximum size of the insert they can carry is small in comparison to the size of the genome. This means that the number of clones necessary to have a reasonably complete genomic library is very large and ordering them is very difficult. Moreover the clones are not always stable and deletions and other rearrangements occur during propagation of the cloned DNA. Cosmids and λ vectors have been most useful for local genomic regions rather than whole genome libraries.

The next type of vector to be developed was the YAC (Box 3.3). YAC libraries have been made that contained very large inserts (up to 1 Mb). This meant that the whole genome could be encompassed in a few thousand colonies, allowing clone maps of the whole genome to be constructed. They have been the workhorse for the construction of the physical maps of the genome described below. However there are several major drawbacks to YAC vectors, the most important of which is a very high rate of chimeric clones. **Bacterial artificial chromosome** (BAC) and **P1-derived artificial chromosome** (PAC) vectors (see Box 3.3) accept inserts in the 100–300 kb range and maintain the topology of inserts much more faithfully than YAC vectors. They will be very important in the construction of high-resolution physical maps of individual chromosomes and the preparation of sequence-ready clones (see Chapter 8).

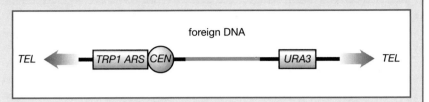

foreign DNA

YACs contain all the elements found in a natural chromosome assembled into a vector that can propagate foreign DNA in a yeast cell. The *ARS* element is an origin of DNA replication, *CEN* is the centromere that ensures segregation to daughter cells and *TEL* is a telomere that protects the chromosome ends. *TRP1* and *URA3* encode enzymes required for the biosynthesis of tryptophan and uracil respectively and are used as selection markers (the host cell is *trp1⁻ ura3⁻*). These essential elements occupy only 6 kb in a circular vector that is maintained in *E. coli*. The two arms of the YAC are purified from this vector and ligated to high molecular weight DNA from the donor genome; the two selection markers ensure that each arm is present. Yeast chromosomes can be up to 1 Mb in size and so DNA in this size range can be carried in a YAC vector. This enormously simplifies the task of constructing and ordering a human genome library because so many fewer clones are required compared with a λ or cosmid library. However, they suffer from three major drawbacks.

1. Up to 60% of the clones are chimeric. It is not clear whether this occurs during the *in vitro* ligation process or *in vivo* by recombination between different vectors after transformation into the yeast cell. The transformation process involves the enzymic removal of the yeast cell wall. This is known to be a mutagenic process and it may stimulate mitotic recombination mechanisms. An alternative method of transformation that keeps the yeast cell intact results in a much lower level of chimeric clones.
2. Clones are unstable in the yeast host, with deletions occurring at a high frequency.
3. There is no easy way to purify the recombinant YAC vector from the remaining yeast chromosomes. It is necessary to fractionate the chromosomes by pulsed field gel electrophoresis and recover the chromosomes from blocks cut out of the agarose slab used for electrophoresis.

Phage P1 is a temperate bacteriophage. It can exist in one of two states in the host cell: in a lytic cycle where it multiplies and eventually lyses the cell, releasing progeny phage particles into the surrounding environment or in a dormant lysogenic state, where it is passively replicated along with the host chromosome. Lysogenic phage are induced to re-enter the lytic cycle by cues such as damage to the host DNA. It has been known for a long time that phage P1 mediates generalised transduction, i.e. it can mediate gene transfer from one *E. coli* cell to another. This formed the basis of one of the original methods used to map the *E. coli* genome. What happens in generalised transduction is that fragments of host

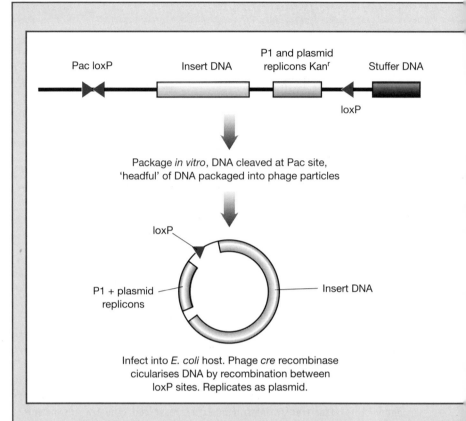

Pac loxP

Insert DNA

P1 and plasmid
replicons Kan[r]

Stuffer DNA

loxP

Package *in vitro*, DNA cleaved at Pac site,
'headful' of DNA packaged into phage particles

loxP

P1 + plasmid
replicons

Insert DNA

Infect into *E. coli* host. Phage *cre* recombinase
cicularises DNA by recombination between
loxP sites. Replicates as plasmid.

DNA are occasionally mistakenly incorporated into the phage head during a lytic cycle. Th
DNA is then injected into a new cell when the phage particle attaches to its surface. Th
ability to transfer non-phage DNA is the basis of PAC vectors. They can accept about 100
of insert DNA and maintain it in low copy number in the host cell as a replicating plasm
when desired the phage lytic system can be induced to increase copy number. The level
chimeric clones in libraries is less than 5%, which is much lower than YACs.

Ordering the clones

Once a complete gene library has been constructed the clones must be
ordered with respect to each other. This is done by detecting sequences present in two clones which indicate that they overlap. This can be done by a
variety of methods, including STS content, hybridisation and fingerprinting.

STS content mapping seeks to identify clones that both contain the
same STS site, indicating that they share an overlapping sequence. The use
of STS content to order clones spanning part of human chromosome 21q is
illustrated in Figure 3.12.

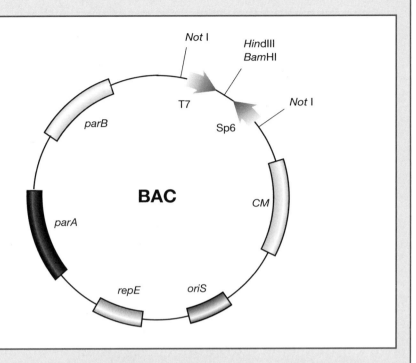

BACs are based on the F factor, the plasmid responsible for conjugation in *E. coli*. The *iS* and *repE* elements mediate replication and *parA* and *parB* maintain copy number at one two per genome. *CM*^r (chloramphenicol resistance) provides a means of selection. Insert NA is cloned into the *Bam*HI or *Hind*III sites and excised using the flanking *Not* I sites. The serts can be transcribed *in vitro* to make RNA using the T7 and Sp6 promoters. BAC vectors n carry up to 300 bp DNA with a high level of stability and low rate of chimeric clones. One sadvantage of BAC vectors is that it is difficult to recover good yields of the recombinant NA from the host because of the low copy number.

If two clones contain sequences in common, they will hybridise to each other. In principle one clone can be used as a probe to identify overlapping clones by using the DNA from one clone as a probe to hybridise to other clones. Hybridisation is more sensitive than fingerprinting or STS content and is often used to detect overlap between adjacent contigs that have been missed by these other methods. It thus provides a method of closing the gaps between adjacent contigs. A problem encountered with hybridisation is a high rate of false-positive results. One reason for this is the presence of repetitive elements in the probe that hybridise to a large number of unrelated clones. One way to ensure that the probe carries only single sequence

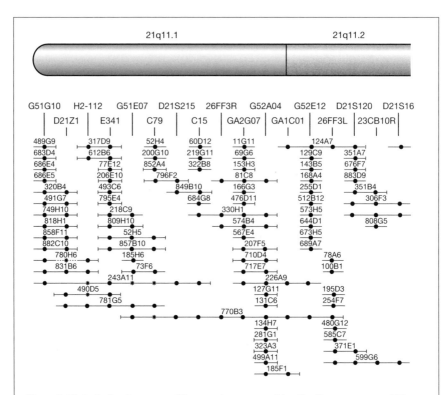

Figure 3.12 Part of a clone map of human chromosome 21q. The figure represents YAC clones, shown as lines, spanning the terminal G-band of the q arm (cytogenetic map is shown at the top). The clones are aligned by the overlap in the STS sites (shown as filled circles) in different clones. The names of the STSs are shown underneath the cytogenetic map in the order revealed by the STS content of the clones. Bars at the end of clone lines means that the clone tested negative against adjacent clones.

DNA is to use PCR to amplify the sequences between adjacent *Alu* elements (see Chapter 2) (Figure 3.13).

Fingerprinting methods aim to identify some property of a DNA sequence that allows it to be recognised where it occurs in two different clones. There are many ingenious ways in which this can be done. Figure 3.14 illustrates a simple method of how this might be achieved.

A clone map of the human genome

The first comprehensive clone map of the human genome was generated by international collaboration. The map consisted of YAC clones ordered into 225 separate contigs. The ordering was achieved by STS content, hybridisation and fingerprinting and was estimated to cover about 75% of the total genome. Only 786 independent loci were fully mapped so the resolution is much lower than the target of the Human Genome Project.

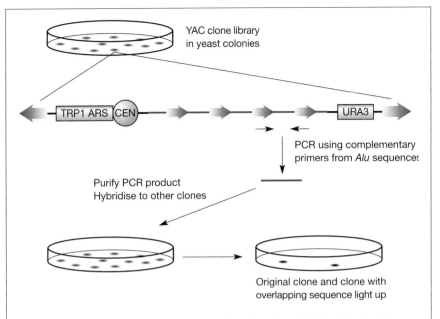

Figure 3.13 Detecting overlap between YAC colonies by colony hybridisation. Each clone contains a large insert of human DNA that has *Alu* repetitive DNA sequences along its length (large blue arrows). Complementary PCR primers (small black arrows) are designed from the *Alu* sequence and are used to amplify unique-sequence DNA located between the *Alu* sequences. The amplified products are used as a hybridisation probe to identify other clones in the library that contain the same sequence and therefore overlap with the first clone.

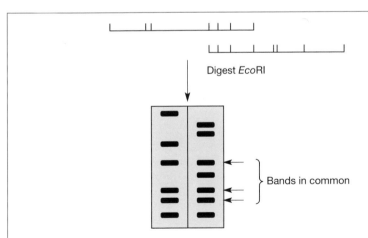

Figure 3.14 Detecting overlap between two overlapping clones by fingerprinting. The clones are digested with a restriction enzyme and the products run on a gel. Clones that overlap will display common bands. It is important to understand that there is no attempt to order the sites within the clones – overlap is indicated by the pattern of bands. Clearly many clones will share bands by chance. The number of apparently common bands expected by chance can be calculated and overlap is only declared when the number of common bands is significantly greater. In practice more complex protocols are used to derive arbitrary patterns that characterise clones.

3.4.2 Radiation hybrid maps

In RH mapping a human cell line is irradiated with X-rays before fusion to hamster cells to make lines of somatic cell hybrids (Figure 3.15). The radiation fragments the human chromosomes so that they cannot be independently maintained in the hybrid unless they become incorporated into the hamster chromosomes. Each hybrid line retains about 20% of the donor human genome in fragments of about 10 Mb. A panel of lines are examined for co-retention of different STS markers. If two markers are co-retained more frequently than would be expected by chance, then this is evidence that they are linked. The distance between markers is estimated from the frequency of co-retention. The map units are **centirays**, which are analogous to centimorgans but depend on radiation dose.

The value of RH maps is that they provide an independent method of mapping STS sites. Because it does not rely on the generation of clones it is not subject to the same systematic errors such as chimeras and deletions that affect clone maps. Moreover, RH maps can detect linkage between STSs over a much longer distance than can be detected by the overlap of clones. They therefore provide a framework to link clone contigs together.

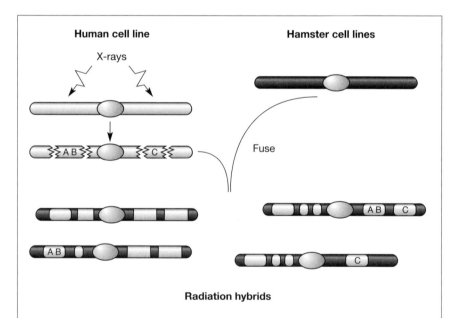

Figure 3.15 Construction of RH maps. Each chromosome represents a complete set of chromosomes in a cell line. A human cell line is irradiated with X-rays, which fragments the chromosomes. The irradiated cells are fused to a hamster cell line. The fragments of human chromosomes become incorporated into the hamster chromosomes. Markers adjacent to each other in the human genome (AB) show the same pattern of retention in the hybrid lines, i.e. they are co-retained. Marker C is distant from A and B and shows a different co-retention pattern compared with either A or B.

3.4.3 Expression maps

ESTs are a special form of STS that are generated from cDNA libraries not genomic sequences. ESTs therefore identify coding regions and may be regarded as gene-based STSs (Figure 3.16). ESTs provide a wealth of information concerning the total number of genes encoded by the human genome, the nature of the proteins encoded and the differences in expression between different tissues.

The human genome is thought to contain between 65 000 and 80 000 genes (see Chapter 2). So far 16 000 of these have been placed on the integrated physical and genetic map screening panels of YAC clones and radiation hybrids for the presence of ESTs. This represents the first step in a programme to map the majority of human genes that will complete the construction of an integrated genetic–physical–expression map.

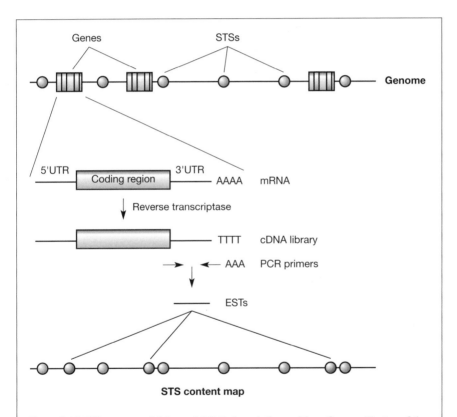

Figure 3.16 ESTs are a special form of STS that mark the position of genes. The top of the diagram shows the genome with isolated genes, each containing introns, shown as blue boxes. The STS content map is superimposed. A cDNA library is constructed from the mRNA derived from the genes. ESTs are prepared from the 3′ untranslated regions (3′ UTR). The 3′ UTR is used because it does not generally contain introns and tends to be unique, even in genes that have related coding sequences. These form STSs which mark the position of genes in the STS content map (filled blue circles).

Physical maps

3.4.4 Long-range restriction maps

Restriction maps have long been an important form of physical map for DNA sequences of up to a few tens of kilobases. The landmarks mapped are the sequences at which different enzymes cut the DNA molecule. Conventional methods of producing restriction maps cannot be used to generate longer maps, which might extend over hundreds of kilobases or even megabases, for two reasons.

1. Commonly used enzymes, such as *Eco*RI or *Bam*HI, cut at 6-bp target sequences. If random sequence DNA is digested with such an enzyme, the size distribution of the fragments produced would have a mode of 4096 bp (4^6). It would be completely impracticable to attempt to order the hundreds of fragments that would be produced if DNA in the megabase size range was digested with such an enzyme.
2. If a way could be found to produce larger fragments, conventional agarose gel electrophoresis (AGE) could not be used to characterise them as it becomes non-linear at large fragment sizes. As the size is increased a point is reached where additional increase in size does not result in a decrease in migration rate. This is because the DNA molecules align along the electric field and migrate end-on through the agarose matrix.

The first of these problems was solved by the discovery of so-called **rare cutting enzymes**. Some examples of these are shown in Table 3.1. These cut less frequently in human DNA than normal enzymes for the following reasons.

Table 3.1 Some examples of enzymes that cut human DNA rarely.

Enzyme	Target sequence
Mlu I	ACGCGT
Not I	GCGGCCGC
Nru I	TCGCGA
Sal I	GTCGAC
Sfi I	GGCCNNNNNGGCC
Xho I	CTCGAG

- Some enzymes such as *Not* I have an 8-bp recognition target. In random sequence DNA this will only occur every 65 kb.

- They contain the dinucleotide CpG. This has two consequences. Firstly the frequency of CpG is depressed in human DNA (see section 2.2.1). Secondly CpG is the target for DNA methylation, which prevents these restriction enzymes from cutting the DNA.

As a consequence the restriction enzyme *Not* I cuts human DNA on average every 500 kb, and its target sites thus provide landmarks that are conveniently spaced for the analysis of human chromosomal regions. Other enzymes, such as *Nru* I and *Sfi* I also produce fragments that are large enough that the target sites are useful landmarks for chromosomal maps. Note that enzymes such as *Xho* I and *Sal* I are comparatively rare cutters with human chromosomal DNA, even though they are commonly used in the construction of conventional restriction maps.

The problem of separating large DNA molecules was solved by a modification of AGE called **pulsed field gel electrophoresis** (PFGE). In PFGE the direction of the electric field is periodically changed. In order to make progress in the new field the DNA molecule must reorientate itself. The time taken for this reorientation is proportional to molecular weight, even with DNA molecules several megabases in size. The exact conditions of electrophoresis can be elaborate and have to be fine tuned for the particular size range employed.

Long-range restriction maps have proved very important during the course of positional cloning. They provide a framework with which to locate the position of clones derived from techniques such as chromosome walking. An example of a long-range restriction map constructed during the hunt for the cystic fibrosis gene is shown in Box 4.2. Another important practical consideration is that the sites of rare cutting enzymes will tend to be clustered in CpG islands that mark the beginning of genes. This allows the islands to be identified in long stretches of DNA.

3.5 An integrated physical and genetic map of the human genome

STSs can be used as a common marker to align genetic, RH and YAC-based clone maps of the human genome. Genetic maps can detect linkage between markers over distances of 30 Mb, while RH mapping can detect linkage over distances of 10 Mb. As a consequence, comprehensive genetic and RH maps of the genome require only a few thousand markers. However the resolution of fine structure mapping is low. In contrast, YAC-based clone maps detect linkage over the 1 Mb base range. Thus a large number of markers are required to provide adequate coverage of the genome, but the fine-structure resolution is high.

As we have seen, clone-based maps suffer from two particular problems: there are always gaps between the contigs, and false linkage between markers may be detected as a result of chimeric clones. RH and genetic maps may be used to provide a framework for clone-based maps, allowing adjacent clone contigs to be positioned with respect to each other and trapping errors that result from chimeric clones. Work is in progress to construct an integrated map that aligns the Genethon genetic map with an RH map and an STS-content map based on YAC clones (Figure 3.17). As of August 1997 this map contained 24 568 STS markers in total, with 10 850 STSs mapped by clone content, 14 665 RH-mapped STSs and 5 264 genetic markers from the Genethon map. For further updates see the Whitehead Institute Web site (see Table 3.2).

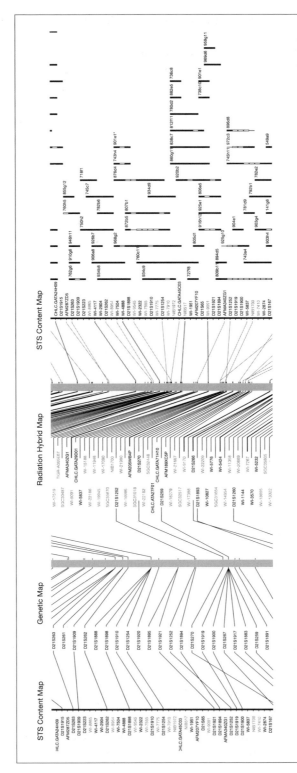

Figure 3.17 A small portion of the integrated map of chromosome 14. The figure follows the same principles as the idealised map shown in Figure 3.9. The STS content map is shown twice, flanking the genetic and RH maps. The lines connect markers found in more than one map. The bars on the right are the overlapping YAC clones which contain the STSs mapped.

As discussed in the preceding section, work is also in progress to plot the position of all human genes using ESTs as a marker. When this is completed, the work of constructing large-scale maps of the human genome will draw to a close. However, work will need to continue to produce maps of individual chromosomes in finer detail as a necessary preliminary to genome sequencing.

3.6 Genome mapping is big science

A traditional view of the way that biological research is carried out would be of a small team of scientists working in a single laboratory. The scale of the effort required to produce whole genome maps has demanded a totally different method of working that is perhaps more familiar to the physical sciences such as particle physics. As an example, consider the following facts concerning the production of the integrated physical and genetic map (see Further reading).

- It involved 15 million separate PCR reactions.

- There were 51 authors to the paper.

- It required the construction of a robotic production line to carry out the PCR tests that was commissioned from a specialist engineering company.

- A full description of the results would have required 900 printed pages. For this reason the detailed results are published electronically and accessed via the World Wide Web (see Table 3.2).

These facts illustrate a number of aspects of how this type of work is carried out.

- Large numbers of personnel are involved. Often several laboratories or even institutions formally collaborate in consortia.

- Automation is essential to allow the large number of tests to be carried out. One value of PCR-based testing is that it readily lends itself to automation.

- Results are published electronically. This allows continuous publication, so that other researchers can exploit the results without the long delays of conventional means of publication. A list of Web sites is shown in Table 3.2. These can be readily accessed to monitor progress of the different mapping projects and to explore the maps in detail. The reader is encouraged to visit these sites.

- PCR-based tests allow researchers to use the sequence of the PCR primers to synthesise their own primers locally and so use the maps in their own particular research.

Table 3.2 Uniform resource locators (URLs) for Web sites of interest to human genome maps.

URL	Comment
http://www.chlc.org/HomePage.html	Cooperative Human Linkage Centre. Contains genetic maps for each human chromosome. Interactive display allows each map to be examined in greater detail
http://www.ncbi.nlm.nih.gov/SCIENCE96/	For each human chromosome shows maps of transcript distribution along with features of interest
http://www-genome.wi.mit.edu/	Homepage of the Whitehead Institute for genomic research. Gives progress bulletins on the integrated physical/genetic/expression map and downloadable graphic files of the chromosome maps
http://www.nhgri.nih.gov/DIR/GTB/CHR7	An electronic version of a paper describing a sequence-ready physical map of chromosome 7 (see Further reading)
http://www-shgc.stanford.edu/	Stanford Human Genome Centre. Physical maps including radiation hybrid maps.
http://www.sanger.ac.uk/HGP	Sequence-ready physical maps of chromosomes X, 1, 6, 20 and 22.

3.7 Summary

- The human genome contains an immense amount of DNA. Maps are necessary to:
  - Locate genes responsible for monogenic and multifactorial disorders.
  - Order individual clones before they are sequenced.

- Maps require landmarks. The most useful form of landmark for genomic mapping is called an STS, which uses PCR to recognise a particular sequence wherever it occurs. The different types of map can all be based on STSs, allowing maps to be aligned.

- Genetic maps are based on recombination at meiosis. They often provide the only connection between biological and medical reality and the underlying genome. A special form of STS based on microsatellites is used as the marker. The definitive genetic map of the human genome has now been produced.

- Physical maps locate features to a particular part of the DNA molecules that make up the genome. There are different types of physical map:

- ○ Clone maps use the overlap between clones in a genomic library to construct groups of clones called contigs, where the spatial relationship of each clone is defined with respect to the other clones in the contig. Comprehensive clone maps are constructed using YACs as vectors because of the large insert size they can carry.
- ○ RH maps plot the position of STS markers by observing how often they are retained together on the same fragment when the genome is fragmented by X-rays and fused to hamster cells to make panels of radiation hybrid lines.
- ○ Expression maps plot the position of genes using a special form of STS called an EST; 16 000 human genes have been mapped in this way (about one-fifth of the total).
- Because the different types of map are based on forms of STS marker, they can be integrated to construct a single STS content map . This is constructed using the STS content of clones to construct contigs, then using genetic and RH maps as a framework to order the contigs and check for errors arising from chimeric clones.

- Mapping the genome is science on a big scale. It involves consortia of research groups, robotic production lines to carry out the tests and electronic publication on the World Wide Web because the data is too copious to publish in printed form.

Further reading

LOD score analysis
RISCH, R. (1992) Genetic linkage: interpreting LOD scores. *Science*, **255**, 803–804.

The genetic map
DIB, C., FAURE, S., FIZAMES, C. *et al.* (1996) A comprehensive genetic map of the human genome. *Nature*, **380**, 152–154.

Physical maps
BOUFFARD, G.G., IDOL, J.R., BRADEN, V.V. *et al.* (1997) A physical map of human chromosome 7: an integrated YAC contig map with an average STS spacing of 79kb. *Genome Research*, **7**, 673–692.

CHUMAKOV, I., RIGAULT, P., GUILLOU, S. *et al.* (1992) Continuum of overlapping clones spanning the entire human Y chromosome 21q. *Nature*, **359**, 380–387.

COHEN, D., CHUMAKOV, I. and WEISSENBACH, J. (1993) A first generation map of the human genome. *Nature*, **366**, 698–701.

FOOTE, S., VOLLRATH, D., HILTON, A. and PAGE, D.C. (1992) The human Y chromosome: overlapping clones spanning the euchromatic region. *Science*, **258**, 60–66.

The integrated map
HUDSON, T.J., STEIN, L.D., GERETY, S.S. *et al.* (1995) An STS-based map of the human genome. *Science*, **270**, 1945–1955.

ESTs and expression maps

ADAMS, M.D., KERLAVAGE, A.R. FLEISCHMANN, R.D. *et al.* (1995) Initial assessment of human gene diversity and expression patterns based upon 83 million nucleotides of cDNA sequence. *Nature,* **377** (Suppl.) 3–17.

SCHULER, G.D., BOGUSKI, M.S., STEWART, E.A. *et al.* (1996) A gene map of the human genome. *Science,* **274**, 540–545.

Sequence tagged sites

ROBERTS, L. (1989) New game plan for genome mapping. *Science,* **245**, 1438–1440.

Radiation hybrid mapping

LEACH, R.J. and O'CONNELL, P. (1997) Mapping of mammalian genomes with radiation (Goss and Harris) hybrids. *Advances in Genetics,* **33**, 63–101.

Commentaries on genomic maps

BOGUSKI, M.S. and SCHULER, G.D. (1995) ESTablishing a human transcript map. *Nature Genetics,* **10**, 369–370.

COX, D.R. and MYERS, R.M. (1996) A map to the future. *Nature Genetics,* **12**, 117–118.

JORDAN, E. and COLLINS, F.S. (1996) A march of the genetic maps. *Nature,* 380, 111–112.

LITTLE, P. (1995) Navigational progress. *Nature,* **377**, 286–287.

SCHMITT, K. and GOODFELLOW, P.N. (1994) Predicting the future. *Nature Genetics,* **7**, 219–220.

Long-range restriction mapping

BARLOW, D.P. and LEHRACH, H. (1987) Genetics by gel electrophoresis: the impact of pulsed field gel electrophoresis on mammalian genetics. *Trends in Genetics,* **3**, 167–171.

Web sites

Visit the Web sites listed in Table 3.2.

CHAPTER FOUR

Molecular analysis of single-gene disorders

Key topics

- Cloning disease genes
- *CFTR* gene
- Dystrophin gene
- Trinucleotide repeat expansions
- Haemoglobinopathies
- Inherited predisposition to cancer
  - ○ Retinoblastoma and the tumour suppressor hypothesis
  - ○ Li–Fraumeni syndrome
  - ○ Hereditary breast cancer: *BRCA1* and *BRCA2*
  - ○ Colorectal cancer: FAP and HNPCC
  - ○ Neurofibromatosis
  - ○ Ataxia telangiectasia

4.1 Introduction

Cloning human genes can be immensely difficult. In some cases it has required large-scale collaborations with many hundreds of scientists and taken years to accomplish. Nevertheless, the potential rewards are equally high and it has transformed research into many inherited diseases. Often, cloning a gene makes it possible for the first time to know about the nature of the protein whose malfunction causes a disease. The type of mutation in affected individuals may be determined, leading to diagnostic tests for the presence of the mutation in families at risk. Finally, it may become possible to cure the disease by **gene therapy**, in which a functioning copy of the gene is introduced to complement the defective gene. This chapter considers how genes are cloned and examines what we learn about the nature of some of the major monogenic diseases by cloning the genes responsible.

and progress towards gene therapy in Chapter 6.

The amino acid sequence of the protein encoded by the gene can be deduced from the nucleotide sequence of the gene. The amino acid sequence may show similarities to other proteins that have been studied either directly in humans or in model organisms such as bacteria, yeast, roundworm, *Drosophila* and mice. This will provide important clues to its cellular function and illuminate the molecular pathology arising from its malfunction. Knowing about the molecular defect helps research to develop better and more rational therapies for the disease.

The nature of mutations can also provide important clues about the protein's function. **Frameshift, nonsense** and **deletion mutations** will be likely to prevent any active protein being made. We may therefore deduce that the defect arises through a lack of function. Lack of function is not the only way in which a gene may be damaged. Sometimes mutations occur that lead to an increase in activity, perhaps through the loss of regulation so that the protein is active in inappropriate circumstances. We shall also see that cloning genes led to the discovery of a totally unexpected and previously unimagined type of mutation called a **trinucleotide repeat expansion.**

Identification of common mutations can be used to develop diagnostic tests for the presence of a damaged gene in families at risk. One mutation may be very common. Sometimes these are **founder mutations,** which can be shown to have occurred in a common ancestor. Diagnostic tests for these mutations are usually straightforward. In other cases, many different mutations may occur and diagnostic tests are more difficult.

4.2 Cloning disease genes

If the protein encoded by a gene has been characterised, a number of methods are available to identify the correct clone in a gene library. Firstly, a partial amino acid sequence may be used to predict the nucleotide sequence of part of the gene. A synthetic oligonucleotide may then be synthesised to use as a hybridisation probe to recognise a clone with cDNA sequences. Alternatively, the protein can be used to raise an antibody. This can be used to screen a **cDNA library** constructed in an expression vector to ensure the encoded protein is expressed in the host. Such methods have been used to clone important genes, such as that encoding human blood clotting factor XIII where previous research had managed to purify some of the protein.

In most cases, however, the protein has not been characterised prior to cloning the gene that encodes it. So how can the gene be cloned in the first place? The only information that may be used to find the gene is its position in the genetic map. This is exploited in an approach known as positional cloning, which has been the method used to clone many genes in the last few years. The database XREFdb (see Table 8.4) lists 84 human disease genes that have been cloned by positional methods.

In positional cloning, families in which the disease is segregating are studied to detect linkage with polymorphic genetic markers. This involves examining a large number of markers to find one where a particular allele and the occurrence of disease are co-inherited in the family. Such linkage suggests that the marker and the disease gene are in close proximity so that the frequency of meiotic recombination has been reduced. The lower the recombination frequency the closer together they are.

Another important strategy for locating the chromosome region of a gene is to look for cytogenetic rearrangements associated with the disease, such as deletions, inversions and translocations. These events require that the chromosomes are physically broken. A breakpoint within a gene will often result in its inactivation. Using Giemsa and Q-bands (see Chapter 2) it is possible to map the breakpoints and therefore the gene affected. A translocation event was very helpful in locating the DMD gene to Xp21 (Figure 4.1).

Linkage analysis or cytogenetic rearrangements narrow down the location of a gene to a particular chromosomal region. But how is the gene itself to be recognised and isolated? A meiotic recombination fraction of 0.01

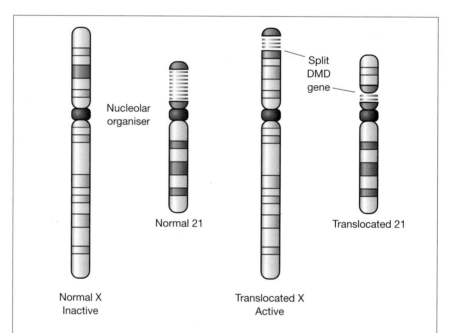

Figure 4.1 Localisation of the DMD gene to Xp21 using a translocation. A female patient suffering from DMD was found to have a translocation between the X chromosome and chromosome 21. The breakpoint on chromosome 21 was within the 28S ribosomal locus. The breakpoint on the X chromosome was at the cytogenetic band Xp21. Although there was still a normal X chromosome, DMD resulted because of unfavourable Lyonisation, i.e. in most of her cells the normal X chromosome was inactivated. The reason for this is not clear, but it could be connected with some form of dosage compensation: if the translocated X chromosome was inactivated there would be two active copies of the X chromosome between Xp21 and the telomere of the p arm (one on the X chromosome and one on the translocated 21 chromosome). Probing a gene library made from this individual with a 28S rDNA probe yielded a junction clone called XJ. This clone contained sequences from the Xp21 region, which proved closely linked to DMD mutations.

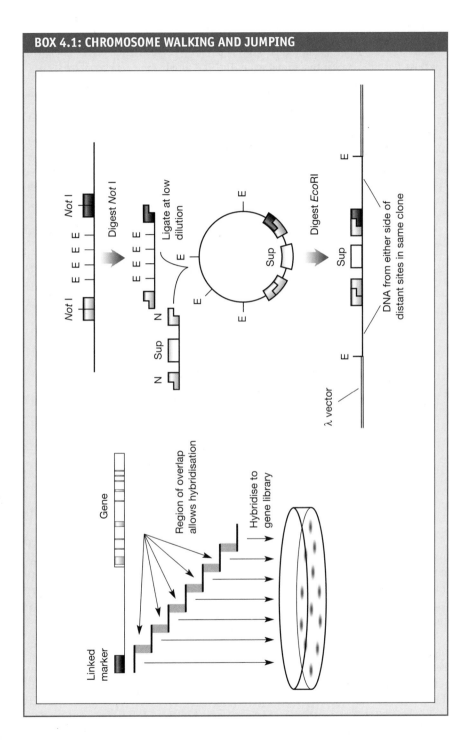

Chromosome walking (shown left) is a way of moving along the chromosome from a linked genetic marker to a target gene. The DNA forming the genetic marker, e.g. a probe detecting an RFLP, is used as a probe in a gene library constructed so that the clones overlap. Unique-sequence DNA isolated from a clone that hybridises with the original clone is used to probe the library again to detect further overlapping clones. The process is repeated many times, each step resulting in a clone that extends towards the target gene. Overall direction of the walk may be monitored using long-range restriction maps made with rare cutting enzymes (see section 3.4.4). Alternatively, new genetic markers may be found in the newly isolated clones. These may be used to measure recombination with mutations in the target gene. Before the use of YAC and BAC vectors, genomic libraries were constructed in λ or cosmid vectors where the insert size was 20 or 40 kb respectively. It required many steps to walk 1 Mb along a chromosome. Chromosome crawling might be a more appropriate term! Sooner or later the walk would be halted because a section of DNA was missing from the library. Such gaps could be bridged by a special technique called chromosome jumping (shown right) which connected adjacent target sites for rare cutting enzymes such as *Not* I. In this technique, chromosomes are digested with *Not* I, which has target sites separated by up to 500 kb of DNA. The fragments are ligated at low concentrations in the presence of higher concentrations of a linker DNA containing a nonsense suppressor and *Not* I compatible ends. The low concentration of fragments favours the formation of very large circles in which the sticky ends from adjacent *Not* I sites are joined together via the linker. The circles are digested with an enzyme such as *Eco*RI, for which there will be many sites between adjacent *Not* I sites, and ligated to a λ vector containing a nonsense mutation in an essential gene. Plaques can only form upon transfection into an *E. coli* host if the linker is present, because it contains the nonsense suppressor. The linker and suppressor are necessary to select the *Eco*RI fragments that contain the *Not* I site and not the other *Eco*RI fragments that will be generated.

(1 cM) corresponds to an average physical separation of 1 Mb, so the linked marker could still be a considerable distance from the gene. The next stage is to isolate overlapping DNA clones from the region. During the cloning of many important disease genes, a process known as chromosome walking was used (Box 4.1). However, the availability of detailed clone maps (discussed in Chapter 3) has made this step increasingly unnecessary.

Identifying the gene

Perhaps the most difficult step in cloning disease genes is identifying the gene responsible in the ordered library collection of clones that result from the chromosome walking. Some genes, such as those for dystrophin, CFTR and factor VIII, are scattered across several hundred kilobases of the chromosome on which they reside. The coding exons sometimes amount to only a few per cent, or even less, of the locus. A number of strategies can be employed to find exons in such genes.

1. Exons encode mRNA, so unique-sequence DNA can be isolated from genomic clones and used to probe **Northern blots** or cDNA libraries to identify regions that are expressed. Any genomic sequences identified in the near vicinity of a gene will be a candidate to be part of the coding sequence. If the **pathophysiology** of the disease indicates that the gene is likely to show tissue-specific transcription, then mRNA from different tissues can be probed to see if the mRNA encoded by the genomic fragment matches the expected pattern.

2. Special techniques such as **exon trapping** can be employed to search for exons that have splice acceptor sites. A number of ways have been developed to do this. Figure 4.2 shows the method used in the search for the HD gene, which played a critical role in its discovery.

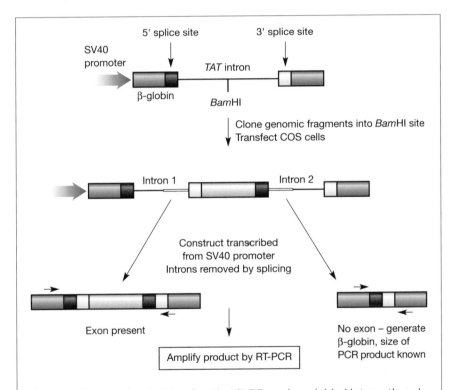

Figure 4.2 Exon trapping. An intron from the HIV *TAT* gene is sandwiched between the end of one exon and the start of the next from the rabbit β-globin gene. Fragments from a genomic region being scanned for exons, approximately 4 kb in size, are shotgun cloned into a *Bam*HI site in the *TAT* intron and transfected into COS cells. The construct is transiently expressed from the SV40 early promoter and the resulting mRNA transcripts amplified by **RT-PCR**, a form of PCR that uses an RNA template, using primers that anneal in the exons. The 5' and 3' splice sites (see section 2.2.1) from the β-globin sequences ensure that the *TAT* intron is removed by splicing; the size of the RT-PCR product will therefore be known. If the genomic insert contains an exon with flanking introns and attendant 5' and 3' splice sites, it will in effect generate two novel introns labelled intron 1 and intron 2. These will be removed by splicing to leave an mRNA transcript containing the exon. RT-PCR now results in a novel product containing the exon.

3. Unique-sequence DNA isolated from genomic clones can be used to probe DNA from other mammals in **Southern blots.** This is known as **zoo blotting.** The rationale is that only protein-coding sequences are likely to be conserved, whereas non-coding sequences are not. DNA from exons will thus cross-hybridise to DNA from other mammals.

4. Some types of mutation, such as deletions and trinucleotide repeat expansions, are relatively easy to identify. If such a mutation co-segregates with the disease, this will map the location of the gene affected.

5. CpG islands (see Chapter 2), which often mark the beginning of genes, can be hunted using enzymes such as *Hpa* II that only cut DNA containing an unmethylated CCGG target site.

6. DNA sequences can be analysed with computer algorithms that detect systematic differences between coding and non-coding DNA. This is more than a question of looking for ORFs, because exons are often not much larger than the length of ORFs that would be expected to occur by chance.

7. Genomic maps are increasingly plotting the nature and position of all expressed sequences (see Chapter 3), thus identifying the location of candidate genes. Sometimes a gene in the region will encode a biologically relevant product. Such a gene would be a candidate for the disease gene; this is known as the **positional candidate** approach. A good example of the positional candidate approach is the cloning of four mismatch DNA repair genes responsible for hereditary non-polyposis colon cancer (see below). This approach is becoming increasingly predominant because of the availability of high-resolution maps.

The application of some of these techniques is illustrated by the cloning of the *CFTR* gene described in Box 4.2.

Once possible genes have been identified, it is necessary to demonstrate which one is actually responsible for the disease. There may be several ways of making a connection. One of the best is to demonstrate co-segregation of mutations in a gene with the occurrence of the disease itself. One problem here is that we can expect variation in base sequence as a result of neutral polymorphisms. Mutations that prevent the protein from being formed, such as nonsense, frameshift and deletion mutations, are unlikely to be neutral to protein function. The co-segregation of these with the disease is a powerful indication that the right gene has been identified. The presence of an expanded trinucleotide repeat is another very useful indicator. Finally, if cells from the affected individual show some phenotype in tissue culture, the ultimate proof is to transfect such cells with the wild-type gene and demonstrate that the defect is corrected.

The first gene to be cloned by positional cloning was that for chronic granulomatous disease in 1986. This was followed by DMD in 1987 and CF in 1989. The early gene hunts were extremely difficult to accomplish, largely because of the low resolution of the genetic maps and the absence of physical and expression maps. The hunt for the CF gene was estimated to have cost $200 million!

HD was the first of the major monogenic disorders to be linked to a genetic marker in 1983. It took a further 10 years and required a large international collaboration to finally isolate the gene. The construction of detailed physical and expression maps of the genome has made this process much easier and arguably most of the important monogenic disorders have now been cloned. The focus is shifting to the more difficult task of cloning polygenic loci that may contribute to the risk of multifactorial disease (see Chapter 5).

4.3 Cystic fibrosis

CF is a common monogenic disorder among northern Europeans. It affects approximately 1 in 2000 people and thus has a carrier frequency of 1 in 22. Its primary symptoms are chronic bacterial infection, inflammation of the lungs and an elevated electrolyte level in sweat. It may also result in pancreatic exocrine insufficiency, obstruction of the bowel in the newborn (**meconium ileus**), diabetes mellitus, liver cirrhosis and male infertility due to CBAVD. The primary defect is decreased chloride ion export and increased sodium ion absorbance leading to insufficient hydration of epithelial surfaces. In the lungs this results in a sticky mucus that cannot be cleared by the action of the cilia that line the epithelia. The failure of this mucociliary clearance mechanism means that foreign particles such as bacteria cannot be removed. For reasons that are not understood at present, infections with *Pseudomonas species* predominate.

4.3.1 CFTR gene and protein

The gene responsible for CF was cloned exclusively from its genetic map position without the aid of cytogenetic rearrangements or knowledge of the protein involved. Box 4.2 describes the process in detail as an example of the difficulties encountered in such an undertaking. The protein encoded by the gene is now known as CFTR.

In Chapter 2 we saw that human genes can be very large and the gene encoding human blood clotting factor VIII was cited as example. The *CFTR* gene is another example of an extremely large gene. It occupies approximately 230 kb of chromosome 7 and consists of 27 exons ranging in size from 38 to 724 bp. Like other large genes, only a small fraction is actually coding sequence. It produces a 6.5 kb transcript that is found in those tissues affected by CF, such as the lungs, pancreas, sweat glands, liver, nasal polyps, salivary glands and colon (Figure 4.3).

Although existing research had suggested that the primary defect in CF was chloride ion transport, prior to cloning the gene there was no direct knowledge of the protein affected. Conceptual translation of the nucleotide sequence showed that the CFTR protein consists of 1480 amino acid residues. It is similar to the ABC (ATP-binding cassette) superfamily of membrane transporters. These proteins are found in all living organisms from bacteria to humans. As their name implies they are

BOX 4.2: THE LONG WALK TO THE CYSTIC FIBROSIS GENE

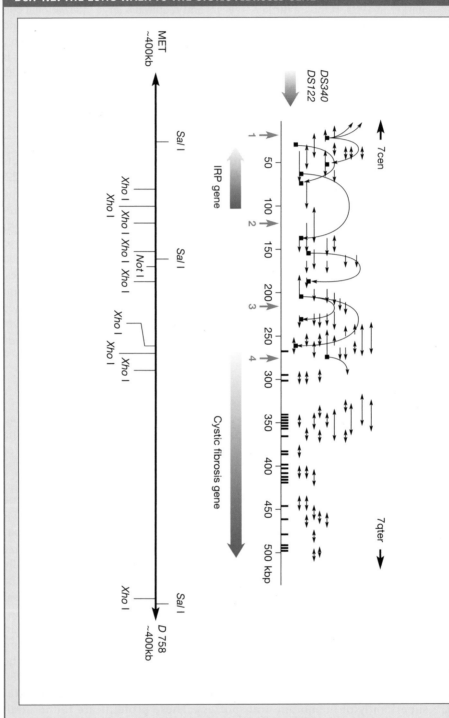

The gene responsible for CF proved exceptionally difficult to clone. This was partly because of the lack of any cytogenetic rearrangements that could be used to map its location accurately and partly because of technical difficulties encountered during cloning. The first step was to map the gene to the chromosomal region 7q21. This was done by observing linkage between CF and two RFLP markers (MET and D7S8). An intensive search for new markers yielded two more, D7S122 and D7S340, that were more closely linked to the *CFTR* gene. Unfortunately, they were only 10 kb apart so effectively they marked the same point on the chromosome. Bidirectional chromosome walks were initiated from the landing places of chromosome jumps from these two markers. In total this involved cloning 280 kb in 49 λ clones and making nine chromosome jumps. In the figure each clone is shown by an arrow, indicating the direction of the step that was made with it. Double-headed arrows indicate clones that were used to walk in both directions. Each jump is shown as an arc. A long-range restriction map was also generated using rare cutting enzymes and pulsed field gel electrophoresis (see section 3.4.4). The map showed that the *CFTR* gene (the name now given to the gene responsible for CF) was about 250 kb in size and was entirely contained within a 380 kb *Sal* I fragment. Progress of the walk was monitored as sites were passed in this long-range restriction map. As well as these clones recovered by walking and jumping (shown in the left-hand part of the diagram), further clones were recovered by probing genomic libraries with cDNA clones (those clones on the right above the *CFTR* gene), so that in total 500 kb of DNA was cloned.

As the walk proceeded, most of the techniques described earlier were applied to identify the *CFTR* gene. Zoo blots proved to be effective and identified four regions whose sequence was conserved in other mammals. Any transcripts encoded by these regions were sought by using the cross-hybridising clones to probe Northern blots and cDNA libraries prepared from tissues affected by CF. Four clones cross-hybridised to other species, shown by the numbered vertical arrows in the figure. Region 1 could be eliminated on genetic grounds. Region 2 corresponded to a gene called *IRP* (INT-related protein) that was known not to be responsible for CF. However, since *IRP* was known to map to the D7S8 side of CF this indicated the direction in which walking and jumping should continue. Although region 3 contained a CpG island that often marks the start of a gene (see Chapter 2), no transcripts could be detected.

Region 4 eventually turned out to be the 5' end of the *CFTR* gene, but this only became apparent after a frustrating series of experiments. The region was identified by two clones called H1.6 and E4.3. The sequences of these clones were rich in undermethylated CpG dinucleotides, suggesting that they were CpG islands. However, they did not contain any ORFs nor did they detect any transcripts in Northern blotting. Eventually, after screening seven different libraries, clone H1.6 hybridised to a single cDNA clone, called 10.1, in a cDNA library made from sweat glands. Clone 10.1 hybridised to a 6.5 kb transcript in a Northern blot and the signal was stronger in cells expected to express the *CFTR* gene. Sequencing showed that the 920 bp insert in clone 10.1 contained a long ORF. Only 113 bp of this sequence overlapped with H1.6. The small overlap explained the difficulty experienced in identifying a transcript. The *CFTR* gene had been found, but only just! Less perseverance or a slightly smaller insert in clone H1.6 and it would have been missed.

Obtaining a full-length cDNA clone again proved more difficult than expected. Clone 10.1 was used to re-probe cDNA libraries; 18 clones were recovered but none contained the full 6.5

kb gene identified in Northern blots. The number of clones recovered was less than expected, and they grew slowly and contained rearrangements. This suggested that clones containing the full-length gene were selected against in the formation or propagation of the cDNA libraries. The consensus sequence of the gene had to be pieced together from the partial sequences of the inserts of these 18 clones. A clone containing the full sequence was then assembled by subcloning from these fragments.

This gene was re-isolated from cDNA libraries prepared from sweat glands of affected and unaffected individuals. A common 3-bp deletion (ΔF508) was exclusively associated with the occurrence of the disease: 145/214 CF chromosomes carried this mutation (68%); in contrast 0/198 normal chromosomes had the mutation ($P<10^{-57.5}$!). The protein encoded by this gene showed strong similarities to the protein superfamily known as the ABC (ATP-binding cassette) membrane transporters. These two observations strongly suggested that the gene responsible for cystic fibrosis had indeed been isolated. The final proof came later with the demonstration that when transfected into epithelial cell cultures derived from CF patients it corrected the chloride ion conductance defect.

The work was carried out as a collaboration between the groups of Lap-Chee Tsui and John Riordan in the Hospital for Sick Children, Toronto, and Francis Collins in Michigan. Lap Chee Tsui's group identified the closely linked D7S122 and D7S340 markers and carried out the chromosome walking and physical mapping in collaboration with the group of Francis Collins, who had developed the chromosome jumping technique. John Riordan's group provided the cDNA libraries and identified the crucial cDNA clones. The outcome was victory in a race with many other groups around the world who were seeking the gene.

responsible for pumping small molecules in and out of cells. Members of this family can be responsible for multidrug resistance in human cells undergoing cancer chemotherapy.

The CFTR protein consists of five domains (Figure 4.3). There are two membrane-spanning domains each consisting of six transmembrane segments. These domains are thought to form a membrane pore through which the chloride ions are transported out of the cell. There are two domains, called nucleotide-binding folds (NBFs), that bind and hydrolyse ATP, essential for the transport process. A domain that has a high proportion of charged amino acid residues separates the two halves of the molecule. This domain is unique to the CFTR protein and has a regulatory function. It contains several sites that are phosphorylated by cAMP-dependent protein kinase (protein kinase A). This phosphorylation is essential for the protein to function and accounts for the prior observation that chloride ion transport is dependent on cAMP-dependent protein kinase.

4.3.2 Mutations affecting the *CFTR* gene

Over 550 different mutations of the *CFTR* gene have been catalogued (Figure 4.3). It is therefore quite astonishing that over two-thirds of CF mutant alleles

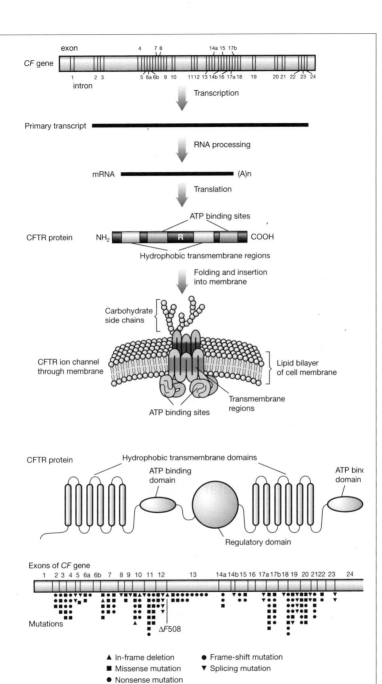

Figure 4.3 Structure of the *CFTR* gene, the transmembrane protein it encodes and the spectrum of mutations that affect its function.

are exactly the same mutation, known as ΔF508. As discussed in Chapter 9, there is evidence that this mutation is extremely ancient, possibly occurring in one of the very first anatomically modern humans that lived in Europe 50 000 years ago and passed on to succeeding generations from this founder. However, this claim is controversial (see Chapter 9).

It has been suggested that carriers of a CF mutant allele enjoy some selective advantage. One possibility is that as CF carriers have a reduced amount of the CFTR protein, there may be a decreased flow of water across the intestinal mucosa. This may make heterozygotes better able to survive intestinal infections that result in infantile diarrhoea. If this is so, then it tells us something about the living conditions of early human populations. Early Europeans must have been exposed to different diseases than other humans as they were the only populations where CF alleles became frequent.

The F508 mutation is a 3-bp deletion that results in the loss of a phenylalanine residue at position 508 in the polypeptide chain in one of the two NBFs. (Δ is a symbol commonly used to signify a deletion and F is the standard single-letter abbreviation for phenylalanine). The CFTR protein undergoes a complex maturation and folding process before the active form is finally located in the cell membrane. ΔF508 polypeptides are defective in this process as they do not undergo an ATP-dependent folding step in the **Golgi apparatus** and are rapidly degraded. However, if the protein does fold, the mutation does not affect its activity. When cells carrying the ΔF508 mutation are cultured at subphysiological temperatures, folding takes place and the protein is active. This provides a target for therapeutic intervention as small drugs may be found that aid the folding of ΔF508 CFTR polypeptides *in vivo*.

4.4 Duchenne muscular dystrophy

DMD is characterised by progressive failure of muscle growth and wasting (dystrophy) leading to weakness, paralysis and respiratory difficulties. In addition the heart, smooth muscle and central and peripheral nervous systems may be affected. It is a sex-linked disease affecting 1 in 3300 boys. Symptoms usually become apparent in boys by the age of 3, they become confined to a wheelchair in their teenage years and usually die by the age of 30. A milder form, called Becker muscular dystrophy (BMD), is caused by mutations that map to the same locus.

4.4.1 DMD gene

The gene was cloned by positional cloning in 1987. Cytogenetic rearrangements resulting in DMD always have one breakpoint in the region Xp21. One rearrangement resulted in a fusion of Xp21 with the 28S ribosomal RNA locus on chromosome 21, which allowed a marker, called XJ1.1, closely linked to DMD to be isolated (see above). Another important strategy was the use of DNA from a boy identified as 'B.B' who suffered from three X-linked disorders: DMD, chronic granulomatous disease and retinitis pigmentosa. Since the genes responsible for these disorders are closely linked, it seemed likely that B.B.'s X chromosome carried a deletion that affected at least part of all three genes. DNA from the region spanned by this deletion was isolated from an XXXXY cell line using a technique called subtractive hybridisation, which preferentially enriched for the region missing in B.B.'s DNA. One of these clones, called pERT87, was shown to be closely linked to the site of DMD mutations. Clones such as pERT87 and XJ1.1 were used to map the genomic locus, including the construction of a long-range restriction map. pERT was also used to probe cDNA libraries, resulting in the isolation of a number of cDNA clones that together spanned the 14-kb mRNA of the locus. These cDNA clones were then used to make a detailed map of the genomic locus including the positions of exons.

The DMD gene is notable for its huge size. It is composed of 79 exons that together with promoter regions, are scattered across 2.4 Mb, about 2% of the whole X chromosome! Transcription of this region is complex, with different cell types using at least seven distinct promoters between them. A full-length 14-kb mRNA is produced from different promoters in muscle cells, cortical cells and Purkinje cells. A shorter transcript is produced by internal promoters in Schwann cells and glial cells, encoding a protein that is correspondingly smaller.

4.4.2 Dystrophin

The protein encoded by the DMD gene, called **dystrophin**, is composed of 3685 amino acid residues. Its role is shown in Figure 4.4.

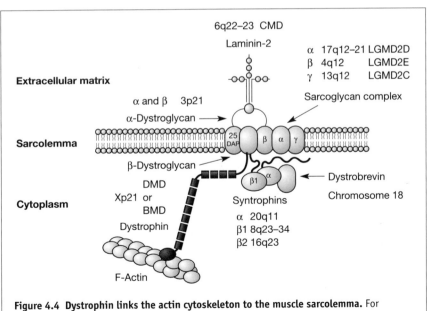

Figure 4.4 Dystrophin links the actin cytoskeleton to the muscle sarcolemma. For explanation see text.

Dystrophin is a long rod-like molecule that links actin fibres in the cortex of muscle cells to the extracellular basal lamina (connective tissue). It does this by forming a bridge between actin and a transmembrane glycoprotein complex called the **dystrophin-associated sarcoglycan complex** (DASC) located in the **sarcolemma**, a membranous sheath that surrounds muscle fibres. DAGC is required for the function and maintenance of muscle cells. It provides a signalling pathway between the connective tissue and the cytoskeleton of muscle cells. In the absence of dystrophin this complex does not form properly. Mutations that affect proteins of the DASC complex also result in a form of muscular dystrophy, called limb girdle muscular dystrophy. In addition, some people suffering from congenital muscular dystrophy (CMD) harbour mutations affecting α-laminin-2, to which DASC is attached outside the cell.

4.4.3 Mutations affecting the DMD gene

DMD mutations arise at a rate of 1×10^{-4} per generation, two orders of magnitude higher than for other X-linked loci. This high rate of mutation is probably due to the target presented by its extreme size. Deletions account for 60% of these mutations. The end-points of the deletions are clustered in two particular introns. The reason for this is not clear, but it could be related to the presence of a mobile genetic element in both introns. Where deletions disrupt the reading frame of the protein, the severe Duchenne form of the disease results. Where the reading frame is preserved, the milder Becker form results. Surprisingly, one mutation has been identified that results in the Becker form, even though the protein is less than half its original size. A similar pattern is seen in small deletions and point mutations: where the mutation allows a full-length protein to form the milder Becker form generally results, but nonsense and frameshift mutations producing truncated proteins cause the severe Duchenne form.

4.5 Trinucleotide repeat expansion mutations

We saw in Chapter 2 that microsatellites consist of tandemly repeated dinucleotide, trinucleotide or tetranucleotide sequences and that the number of repeats is polymorphic. The ability to clone genes and to analyse the mutations responsible for genetic disease led to the unexpected discovery that polymorphic trinucleotide repeats can dramatically increase in length to form a novel type of mutation called a **trinucleotide repeat expansion** (TRE). A total of 11 such mutations have been identified (Table 4.1). In TRE mutations the number of repeats is unstable in meiosis, in some cases even in mitosis, usually causing a further increase in repeat length. For this reason they are sometimes called **dynamic mutations.** Dynamic mutations are responsible for a phenomenon called **anticipation**, where either the severity or the degree of penetrance of a disease apparently increases with each succeeding generation.

Table 4.1 Disorders caused by trinucleotide repeat expansions. 105

Fragile sites caused by CCG expansion in 5' UTR
FRAXA (fragile-X syndrome)
FRAXE (mental retardation)
FRAXF
FRA16A
FRA11B (Jacobsen syndrome)

CTG expansion in 3' UTR
Myotonic dystrophy

Neurodegenerative disorders caused by CAG expansions in coding regions
Huntington's disease
Spinal and bulbar muscular atrophy
Spinocerebellar ataxia
Dentatorubral–pallidoluysian atrophy
Machado–Joseph disease

TREs can broadly be divided into two classes. In one class, exemplified by fragile-X syndrome, the trinucleotide concerned is CCG, which expands in the **untranslated regions** (UTRs) of genes. This typically results in chromosome fragile sites: cytogenetically visible, non-staining gaps in metaphase chromosome spreads. In the other class, the trinucleotide concerned is CAG, which occurs in a number of neurodegenerative disorders such as HD. In this class, the number of expanded repeats is much lower and occurs within the coding region of the gene affected. One TRE mutation that does not easily fit in this classification is responsible for MD. In this case the trinucleotide concerned is CTG and the expansion site is located in the 3' UTR of the gene that is probably affected.

4.5.1 Fragile sites caused by CCG expansions

In TRE mutations involving CCG the normal repeat number of ~30 increases to several hundred or even thousands. It results in **fragile sites** in chromosomes. These are non-staining gaps in chromosomes visible in metaphase spreads of cells cultured under conditions such as folate deprivation or chemical inhibition of DNA synthesis. Five loci involved in fragile sites have been cloned (Table 4.1). All have been shown to involve expansions of a CCG repeat. Some of these fragile sites have no apparent phenotypic consequences. However, one of them, FRAXA, results in fragile-X syndrome, characterised by moderate to severe mental retardation. Two other fragile sites, FRAXE and FRA11B, are also associated with mental retardation.

Fragile-X syndrome is the commonest cause of congenital mental retardation after Down's syndrome, affecting 1 in 2000 children. There are a number of unusual features in its pattern of inheritance.

1. It shows incomplete penetrance as, on average, 20% of boys who are known to carry the allele, because they transmit the disorder to subsequent generations, are not affected. These boys are classified as normal transmitting males. The degree of penetrance increases with each succeeding generation in a pedigree. This phenomenon is called anticipation and, in one form or another, is characteristic of TRE mutations.
2. It shows at least partial dominance because 30% of female heterozygotes are also affected. These affected females always inherit the fragile site from their mothers, never from their fathers.
3. In males the disorder is never caused by a new mutation during gametogenesis in their mothers.

FRAXA, the fragile site associated with fragile-X syndrome, is located at Xq27.3. The gene affected is called *FMR1* (fragile-X mental retardation 1). The expansion affects a polymorphic site in which the number of CCG repeats normally ranges from 6 to 52. Individuals affected by the disease are said to carry a full mutation when the number of repeats increases to 230–1000 copies. The expansion site is located upstream of the coding region near to a CpG island. The full mutation induces **DNA methylation,** which extends into this CpG island and consequently inactivates the *FMR1* gene. Some boys, who suffer from a severe form of the disorder, have been found to have deletions of the *FMR1* gene. This confirms that the expansion does indeed cause loss of gene function and is not acting in some other way to change the activity of the *FMR1* gene product. The role of the FMR1 protein has not yet been fully elucidated. It contains an amino acid sequence that is found within proteins known to bind RNA. The FMR1 protein binds specifically to approximately 4% of mRNA molecules found within brain cells, so it may have a role in controlling translation.

The number of repeats at the expansion site can show intergenerational instability, which accounts for the unusual pattern of inheritance. Normal transmitting males harbour **premutations** in which the number of repeats ranges from 60 to 230. These can expand into full mutations during meiosis in a carrier female but not in a normal transmitting male. The frequency with which the premutation expands into a full mutation is dependent upon its length, which can only increase in a female carrier. With each passage of a premutation through a carrier female its length may increase, consequently raising the chance of expansion into a full mutation in the next female carrier in the pedigree. This explains the phenomenon of anticipation. The requirement for a premutation to be expanded by passage through a carrier female explains why the disorder is never caused in boys by a new mutation during gametogenesis in their mothers. The observation that expansion can only occur in females explains why affected females always receive the mutant allele from their mothers. Both males and females bearing the full mutation show somatic instability where the repeat length and degree of methylation can show wide variation in different cells.

A total of five neurodegenerative disorders have been shown to be caused by expansion of polymorphic CAG repeats located within the coding region of the gene affected (Table 4.1). All of the disorders are late onset, autosomal dominant and progressive in nature. They are all characterised by similar symptoms: ataxia, chorea, dementia and sometimes psychosis. The effect of each disease is limited to a different group of cells in the brain, yet each gene is expressed in many different classes of brain cells; there is currently no explanation for this apparently paradoxical observation. CAG encodes glutamine, so the expansion mutation will cause an increase in a polyglutamine tract in the encoded proteins. Since these are all dominant mutations this must in some way alter the activity of the protein rather than reduce its function.

The best-known and most common of this group of diseases is HD, which affects 1 in 10 000 Caucasians. The symptoms are generally manifested in mid life, but it can show juvenile onset in which case the course of the disease is more rapid and the symptoms more severe. A particularly distressing aspect of HD is that, as a late-onset, autosomal dominant disease, sufferers will normally have had children before the onset of the disease. Each of these children will have a 50% chance of being affected.

In 1983, HD was one of the first human genes to be mapped by linkage to polymorphic DNA markers. However, the gene responsible was only cloned in 1993 after an exceptionally prolonged and difficult gene hunt. Two strategies finally led to the gene.

1. Linkage disequilibrium (see section 5.5.2) showed that although multiple mutations have occurred to cause HD, one-third of HD chromosomes probably descend from the same ancestral chromosome. The haplotype analysis narrowed the location of the gene to a 500-kb region.
2. Exon trapping (see Figure 4.2) was used to identify every transcript that arose from the region identified by the linkage disequilibrium analysis. In HD patients the gene encoding one of these was affected by an expansion of a trinucleotide repeat $(CAG)_{\sim 15}$.

The gene affected, known as *IT15*, is located near the telomere on chromosome 4. It encodes a protein of unknown function called **huntingtin**. The coding region contains a polymorphic CAG repeat that expands in those affected by the disease (Figure 4.5). The normal range of repeat lengths is 15–35; in affected individuals this increases to 36–121. It is remarkable that the lowest number of repeats that causes the disease is only one more than the highest found in normal individuals, although it should be said that repeats in the 30–40 range are extremely rare. Apparently increasing the polyglutamine tract from 35 to 36 residues is enough to trigger the onset of the disorder.

HD does not show anticipation in the formal sense, but there are some related phenomena. Longer repeat lengths are associated with juvenile onset and more severe symptoms. These are generally transmitted by affected fathers, suggesting that expansion takes place during male gameto-

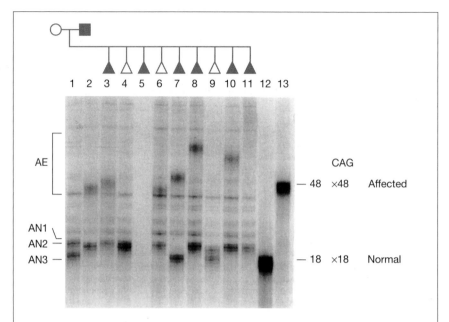

Figure 4.5 Trinucleotide repeat expansion in HD. The figure shows a large family in which the disease is segregating. The region surrounding the CAG repeat site has been amplified using PCR. All individuals in the pedigree have a normal allele with the CAG trinucleotide repeated about 18 times. Affected individuals also have a mutant allele where the repeat number has increased. Lanes 12 and 13 show reference cDNA clones where the repeat length is 18 (wild type) or 48 (mutant). The PCR reaction in lane 5 failed. Individual number 11 apparently only has a normal allele. Subsequent analysis showed that this individual carried an expansion so large that the PCR reaction failed. As a consequence of the very large repeat, the disease was particularly severe and showed juvenile onset.

genesis. This is reminiscent of fragile-X syndrome, where the expansion can only take place in the female germline. Age of onset is also inversely correlated with repeat length.

4.5.3 Myotonic dystrophy

MD (also known as **dystrophia myotonica**) is an autosomal dominant neuromuscular disorder. It affects 1 in 8500 individuals, making it the most common neuromuscular disorder showing adult onset. It also shows a range of symptoms from cataracts and frontal baldness in mildly affected individuals to multisystem failure in severely affected individuals with symptoms including muscular dystrophy, myotonia (an inability to relax muscles after voluntary contraction), endocrine dysfunction, mental retardation, cardiac abnormalities and testicular atrophy in males. It also shows a variable pattern of onset, ranging from a congenital condition with severe symptoms to a classic adult-onset form and a mild, late-onset form. The disorder shows anticipation, with increasing severity and decreasing age of onset with each succeeding generation. Often the symptoms are so mild in the grandparents or parents that the disease had not been diagnosed prior to its occurrence in a severe form in the grandchildren.

The disorder is caused by the expansion of a CTG repeat located in the 3' UTR of a gene known as the **dystrophia myotonica protein kinase** (*DMPK*). The repeat is polymorphic in normal populations with repeat numbers in the range 5–35. This expands in affected individuals to repeat numbers in the range 50 to >3000. There is an approximate correlation between repeat length and severity and age of onset. The number of repeats increases with each succeeding generation, explaining the observed antici-pation. Severely affected congenital cases show the highest number of repeats. Mildly affected or asymptomatic forebears of more severely affected individuals will have between 50 and 80 repeats. The expansion is not sex-limited as it has been observed in parents of both sexes.

There have been conflicting reports as to whether the expansion increases or decreases *DMPK* expression, so it is not clear whether the dis-order results from a gain or loss of gene function. Moreover, neither *DMPK* knockout mice nor mice that overexpress a human *DMPK* transgene show full symptoms of the disease; thus there is still some doubt as to whether the *DMPK* gene is solely responsible for the disorder. The expansion occurs near the CpG island for a second gene called **DM locus associated homeo-domain protein** (*DMAHP*), which may also play a role in the disease as its transcript level is reduced by the expansion allele and its pattern overlaps substantially with that of the *DMPK* gene in mice.

4.6 Haemoglobinopathies

Haemoglobin is the oxygen-carrying molecule in the blood. It forms 90% of the protein content of red blood cells and is by far the most abundant blood pro-tein. It is a tetramer consisting of two α-globin and two β-globin polypeptide chains (α_2 or β_2) each complexed with a molecule of iron-containing haem. It binds oxygen cooperatively, which is important to its physiological function of binding oxygen at the high levels of dissolved oxygen found in the lungs and releasing it at the lower levels found in body tissues. Homotetramers of the same globin type (α_4 or β_4) show hyperbolic oxygen binding and are unable to release oxygen under physiological conditions.

The haemoglobinopathies are a group of diseases that affect haemoglo-bin production or function leading to anaemia. Genetic defects directly affecting the globin genes can be divided into two classes of almost equal prevalence: sickle cell disease and the thalassaemias. As well as sickle cell disease and the thalassaemias, glucose 6-phosphate dehydrogenase defi-ciency is another common hereditary disease that can affect haemoglobin. This results in favism (haemolysis after eating beans), anaemia in response to certain drugs and neonatal jaundice.

In sickle cell disease, β-globin is produced in normal quantities but an amino acid substitution causes it to precipitate into inactive aggregates within red blood cells, resulting in the characteristic sickle morphology. In the thalas-saemias, one of the globin chains is reduced or absent. This results in an excess of the other type, which is responsible for the clinical manifestations. β Thalassaemias describe disorders in which β-globin is reduced or absent; α thalassaemias describe corresponding disorders in which α-globin is reduced or absent. A complete absence of α-globin (α^0 thalassaemia) causes **hydrops fetalis,** which is fatal at birth due to oxygen starvation. Sickle cell anaemia, milder forms of α thalassaemia and the β thalassaemias show a wide range of clinical severity that may be treatable by regular blood transfusions.

Collectively, the haemoglobinopathies are the most common form of disease traceable to a monogenic defect. Approximately 7% of the world's population are thought to be a carrier for one of the disorders. Sickle cell anaemia is most common in Africa where it affects 100 000 infants each year. There are also significant numbers of cases in other regions where there are populations of African descent, such as the USA (1500 new cases annually), the Caribbean (700 new cases annually) and the UK (140 cases annually). The thalassaemias are especially prevalent in South East Asia. In Thailand, for example, 500 000 suffer from some form of hereditary anaemia. β Thalassaemia is also common in Mediterranean peoples. Some Mediterranean islands have a particularly high rate: the carrier frequency is 20% in parts of Sardinia and Cyprus, resulting in a population incidence of 1%.

High frequencies of the haemoglobinopathies correlate closely with the worldwide incidence of malaria, indicating how they have arisen. Heterozygotes are more resistant to infection by *Plasmodium falciparum,* the causative agent of the most severe form of malaria. This leads to a polymorphism, where selection for mutant alleles when carried by heterozygotes is balanced by selection against the same alleles when carried by homozygotes.

4.6.1 Globin genes

Both α-globin and β-globin are encoded by a family of related genes that are expressed at different times during development. The α-globin gene cluster occupies 30 kb of chromosome 16 (Figure 4.6). There are two α-globin genes, α_1 and α_2, encoding identical polypeptides and differing only in small changes in non-coding parts of the gene, such as the 3' UTR and introns. The two genes are thought to have arisen by duplication and this is reflected by blocks of homology in the surrounding regions. As well as the α-globin genes, there is a related α-globin-like gene called ζ_2-globin which is expressed in early embryonic development (see below). In addition there are pseudogenes (see Chapter 2) of both α-globin and ζ-globin called $\psi\alpha_1$ and $\psi\zeta_2$ respectively. Finally, there is a gene that resembles α-globin called θ-globin, the status of which is uncertain. It is transcribed, but the resulting protein is apparently not incorporated into haemoglobin. It may be a pseudogene in the early stages of evolution.

The β-globin gene cluster is 70 kb in size and located on chromosome 11 (Figure 4.7). It consists of five genes and one pseudogene. There are two similar adult forms, β-globin itself and δ-globin, which constitutes about 6% of adult β-globin-like chains. During the foetal stage of development Aγ and Gγ are expressed; they encode β-globin-like proteins that differ only at posi-

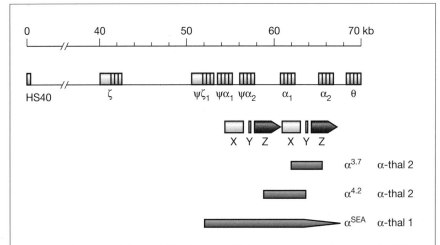

Figure 4.6 Structure of the α-globin cluster and deletions causing α thalassaemia. X, Y and Z are blocks of sequence homology repeated in the regions surrounding α₁ and α₂ genes. The blue bars show the extent of deletions resulting in α thalassaemia. The figures in superscripts show the size of the deleted regions. SEA, South East Asia.

tion 136, which is alanine in Aγ and glycine in Gγ. Finally, ε-globin is expressed in early embryonic development.

As a result of the changing pattern of expression during development, a variety of haemoglobin molecules appear that differ in the composition of the haemoglobin subunits (Table 4.2). These different haemoglobins meet the changing oxygen transport requirements at different developmental stages.

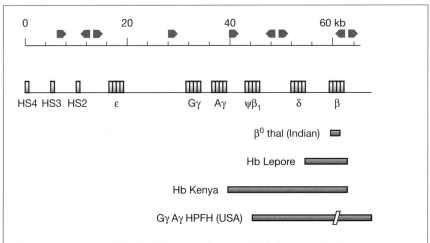

Figure 4.7 Structure of the β-globin gene cluster and deletions causing β-thalassaemia. The solid bars indicate the extent of deletions found in particular deletion mutations. HS4, HS3 and HS2 are locus control elements. The arrows above the scale indicate the location and orientation of *Alu* elements.

Table 4.2 Haemoglobins made at different times during development.

| | β-globins | | | | |
| | Embryonic | | Foetal | Adult | |
α-globins	ε	γ	γ	δ	β
ζ	ζ₂ε₂	ζ₂γ₂			
	Hb Gower1	Hb Portland			
α			α₂γ₂	α₂δ₂	α₂β₂
			Hb F	Hb A₂	Hb A

During development the pattern of expression of both the α-globin and β-globin clusters is under the control of locus control regions (LCRs). HS2–HS4 control the expression of β-globin and HS40 controls the expression of α-globin (see Figures 4.6 and 4.7). LCRs are enhancers that bind erythroid-specifc transcription factors. Without the LCR there is very little expression of the genes, but exactly how they act is still unclear. One striking fact about both clusters is that the physical arrangements of the genes reflect the order with which they are expressed during development, but the significance of this is still unknown. The LCRs map a considerable distance upstream of the clusters. For example, HS40 is located over 60 kb upstream of the α_1 and α_2 genes. Thus they are an extreme example of the way that enhancers can act at distance to activate transcription of the gene they control.

4.6.2 α Thalassaemia

α Thalassaemias are caused by deletions of one or both α-globin genes. An α-thalassaemia 1 (α-thal 1) chromosome has one remaining gene while an α-thalassaemia 2 (α-thal 2)chromosome has both genes deleted. These chromosomes can combine in different ways to result in the presence of zero, one, two or three α-globin genes (Figure 4.8). The severity of the resulting thalassaemia is determined by the number of remaining copies.

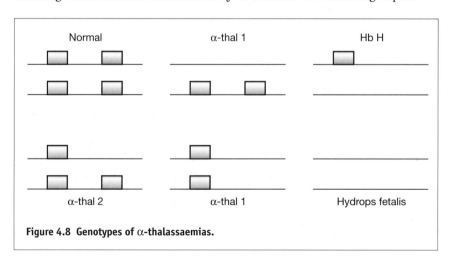

Figure 4.8 Genotypes of α-thalassaemias.

- Individuals with three α-globin genes (α-thal 2) are asymptomatic carriers.

- Two functioning α-globin genes (α-thal 1) results in mild anaemia.

- One functioning α-globin gene results in moderate anaemia with inclusions in red cells consisting of β_4 tetramers known as Hb H.

- Individuals with no α-globin genes die shortly after birth from a condition known as hydrops fetalis. The primary haemoglobin is Hb Barts (γ_4), which has such a high oxygen affinity that it fails to release any oxygen in the tissues.

The deletions apparently result from unequal crossing-over between the duplicated segments which originally gave rise to the two α-globin genes (Figure 4.9). Reciprocal chromosomes with three α–globin genes have been observed.

4.6.3 β Thalassaemia

The severest form of β thalassaemia is β^0 thalassaemia where there is a complete lack of β-globin chains in adult life. The resulting excess of α-globin chains precipitate, leading to red blood cell damage and also inhibiting the production of new red blood cells (**erythropoiesis**). In β^+ thalassaemia there is a large reduction but not complete absence of β-globin synthesis. A relatively benign form of β thalassaemia is hereditary persistence of foetal haemoglobin (HPFH), which occurs when the absence of β-globin is compensated by the continued synthesis of the foetal form. Sometimes, synthesis of foetal haemoglobin persists, but at insufficient levels to compensate for the loss of β-globin; this condition is known as δβ thalassaemia because there is also an absence of δ-globin.

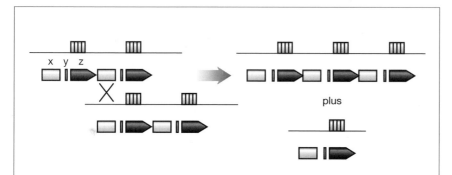

Figure 4.9 Unequal crossing-over results in a deletion of an α-globin gene. Unequal crossing-over between X regions produces one chromosome with a triplicated α-globin locus and a chromosome containing only one. In this example, the resulting deletion would be 4.2 kb in size, as observed in some α-thal 2 chromosomes (Figure 4.6). Chromosomes carrying three α-globin loci have been observed in populations where α-thal 1 is common.

Virtually every type of point mutation that could lead to a reduction or absence of β-globin synthesis has actually been observed in cases of β^0 and β^+ thalassaemia. These include:

- a deletion of the whole β-globin gene;

- promoter mutations preventing or reducing transcription;

- mutations of splice sites preventing the removal of introns;

- mutation of the poly-A acceptor site affecting RNA processing;

- nonsense and frameshift mutations causing termination of translation.

Many deletions affecting the β-globin cluster have been analysed; these are represented in Figure 4.7. One of them, Hb Indian, is the deletion affecting the β-globin gene that results in β^0 thalassaemia. Some of the others are apparently the result of unequal crossing-over between two β-globin genes, resulting in their fusion and the deletion of intervening DNA. For example, Hb Lepore results from a deletion that fuses the 5' portion of δ-globin with the 3' portion of β-globin. The reciprocal product of this recombination event, known as Hb anti-Lepore, has been observed.

4.7 Inherited predisposition to cancer

All cancers can be said to be genetic diseases, in that they arise through damage to 100 or so cellular **oncogenes** which, in their unmutated state, normally cooperate to maintain control over cell proliferation (see Box 4.3). The controls are complex so that, for the most part, failure of one component is compensated by the action of a different part of the control network. Before a cell is completely liberated from proliferation controls, multiple parts of the control system must be damaged. The origin of a tumour, and its progression to an invasive malignant state, is thus a multi-stage process: successive mutations to different oncogenes are necessary. For this reason the risk of cancer increases progressively with age as the mutations accumulate.

The mutations may be caused by endogenous cellular processes or they may be inflicted by environmental mutagens. Rarely, individuals are born with germline mutations affecting key components of the regulatory network. These are responsible for hereditary predispositions to cancer. Many of the products of these genes have also been found to be mutated in sporadic cancers and the study of the genes affected by these mutations has provided critical insights in understanding the mechanisms that normally control cell division as well as allowing diagnostic tests for families at risk.

Products of oncogenes operate at different levels. Some function in signal transduction pathways, ensuring that cells only divide when positively stimulated to do so by external **growth factors** collectively known as **mitogens**. Others may encode cellular receptors of the external signal, components of the intracellular signalling apparatus or nuclear transcription factors that change the pattern of gene expression. Mutations in these pathways are often dominant at a cellular level: they change the pattern of activity or result in hyperactivity of the gene product so that the signalling pathway is activated inappropriately. One of the most commonly occurring types of mutation in this category are mutations to members of the **ras** family, which act at the interface of growth factor receptors and intracellular signalling pathways.

Other types of cell division regulators act to inhibit cell cycle progression. They provide mechanisms, called **checkpoints**, that prevent cell division when it would be inappropriate, for example they may halt the cell division cycle to allow damage to DNA to be repaired. In presence of severe damage to DNA they even direct the cell to commit suicide, a process known as **apoptosis**. The way in which they operate is described in Box 4.3. Finally, other types of gene product are involved in the maintenance of genome

BOX 4.3: THE CONTROL OF CELL DIVISION

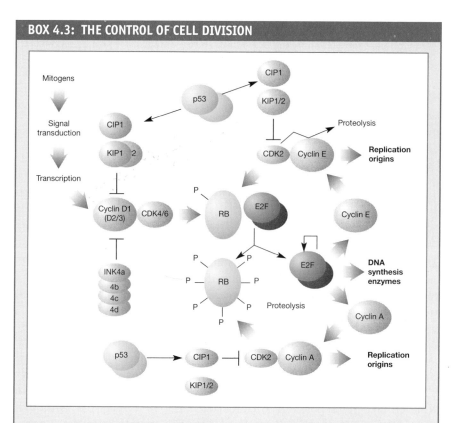

Progress through the cell cycle is controlled by the **cell cycle engine**, which consists of a family of **protein kinases** called cyclin-dependent protein kinases (CDKs). As their name implies, these kinases must be associated with a cyclin partner to be active. The periodic appearance and disappearance of different cyclin families controls the action of the different CDKs, which in turn regulates the passage of cells to division. Cells are normally only able to divide in the presence of mitogens; in the absence of mitogens cells remain in a quiescent state with unreplicated DNA (a state sometimes called G_0). Stimulation of a quiescent cell results in the rise of the cyclin D family of proteins, of which there are three members, cyclin D1, D2 and D3, which differentially respond to various signals in different tissues. Cyclin D molecules associate with CDK4 or CDK6. This complex phosphorylates the retinoblastoma protein (RB), causing it to release a heterodynamic transcription factor called E2F. Free E2F activates transcription of genes required for DNA synthesis. E2F may also activate its own transcription and that of cyclin E and cyclin A. Cyclin E expression is periodic, normally peaking in late **G_1**. The catalytic partner of cyclin E is CDK2 which also phosphorylates RB, making this process independent of cyclin D/CDK4/6 and hence the presence of mitogens. Cyclin A normally appears in **S phase** and levels remain high throughout the rest of the cycle. It associates with CDK2 in early S phase and with CDC2 in **G_2**. Cyclin E/CDK2 and cyclin A/CDK2 carry out key phosphorylation steps that activate replication origins and so trigger DNA replication. Once this occurs, E2F is inactivated through phosphorylation by cyclin A/CDK2, and cyclin E is destroyed by proteolysis to prevent further rounds of DNA synthesis. Cyclin E may be targeted for proteolysis by phosphorylation by its own CDK2 partner.

A key part in the regulation of this process are two families of CDK inhibitors called the INK4 family (four members) and the CIP1/KIP family. The INK4 proteins specifically inhibit cyclin D/CDK4. Mutations that inactivate INK4 or cause amplification of the cyclin D gene have been found in a wide variety of commonly occurring cancers. CIP1 is induced by p53, which mediates the DNA damage checkpoint pathway, pausing the cell cycle to allow DNA repair. In the presence of severe damage p53 provokes apoptosis. DNA damage is sensed by p53 through a pathway that includes the AT protein, which is inactivated in ataxia telangiectasia.

integrity and DNA repair. Failure of checkpoint and DNA repair functions allows other genetic changes to accumulate, which may further incapacitate the mechanisms that control cell proliferation. Oncogenes therefore occupy a key position in the regulatory network and their failure has been implicated in virtually every type of cancer.

Mutation to genes encoding checkpoint and repair functions are generally recessive at a cellular level because inactivation of one copy of the gene can be compensated by the function of the other copy. Such genes are known as **tumour suppressors** or **anti-oncogenes** because they act to prevent tumours from occurring. Many of the germline mutations that give rise to hereditary predisposition to cancer are tumour suppressors. Although they are recessive at the cellular level, germline mutations to these genes result in a predisposition to cancer that is inherited in an autosomal dominant pattern. This comes about because a cell in which one copy of the gene is inactivated by a germline mutation requires one further mutation event to inactivate the other. The chances that this will occur during somatic development are high and tumours arise at a high frequency. In contrast two independent mutation events are required in somatic cells not carrying a germline mutation; such events will occur and result in sporadic cancers, but at a lower frequency.

Germline mutations to tumour suppressor genes commonly result in predisposition to tumours in diverse types of tissue. Sometimes, as in the case of retinoblastoma, one tissue predominates, presumably because of additional controls that protect cells in other tissues. However, in such cases tumours do occur in other tissues. In other cases, such as Li–Fraumeni syndrome, the predisposition is to multiple tumour types.

4.7.1 Retinoblastoma

The study of **retinoblastoma** allowed Alfred Knudson to formulate the tumour suppressor paradigm described above, which has been central to the study of hereditary cancers. Retinoblastoma is a rare childhood cancer of the retina, affecting about 5 in 100 000. About 40% of the cases show an autosomal dominant pattern of inheritance while the remaining 60% are sporadic. Several features of the inherited form distinguish it from the sporadic form. The penetrance of the hereditary form is 95%, the mean number of tumours is three and tumours in both eyes are common. Survivors are susceptible to other tumours, particularly osteosarcoma and soft-tissue sarcomas, at a rate of about 15% by age 30. In contrast the children with the sporadic form will normally suffer from a single tumour in one eye and are not susceptible to other types of cancer.

Knudson proposed that the hereditary form occurs when one copy of the gene involved, called *RB1*, is inactivated in the germline and the other is inactivated by mutation during somatic development. He showed that the known rate of mutation at the *RB* locus, the number of target embryonic **retinoblasts** (the embryonic precursor cells), the penetrance of the hereditary form and population frequency of the sporadic form were all consistent with such a hypothesis. Multiple tumours occur in the hereditary form because of the high probability that more than one retinoblast will be affected by a second mutation hit. The increased occurrence of other tumour types reflects a common role for the *RB1* gene product in proliferation control in other tissues. In contrast to the hereditary form, multiple tumours are rare in sporadic cases because it is extremely unlikely that more than one retinoblast will be independently affected by two hits.

This model predicts that when the second copy of the *RB1* gene is examined it will be found to have been inactivated. This can happen in a variety of ways: (i) point mutation, (ii) deletion, (iii) chromosome non-disjunction or (iv) mitotic crossing-over. Three of these mechanisms will result in tumour cells becoming homozygous for markers near the *RB1* locus on the chromosome bearing the germline mutation (Figure 4.10). Such a situation is known as **loss of heterozygosity** (LOH).

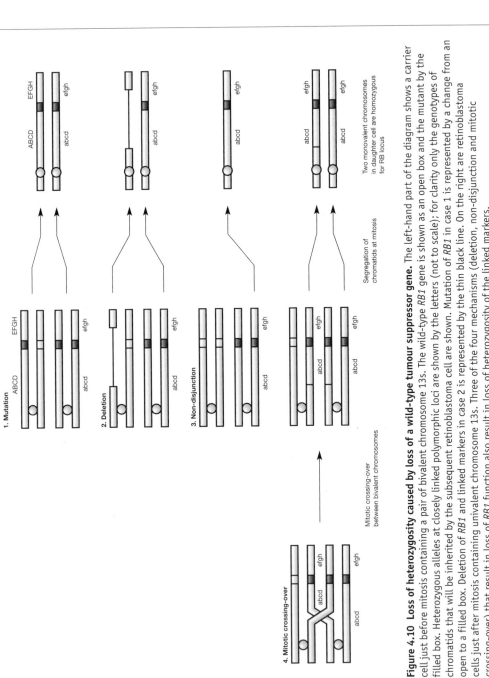

Figure 4.10 Loss of heterozygosity caused by loss of a wild-type tumour suppressor gene. The left-hand part of the diagram shows a carrier cell just before mitosis containing a pair of bivalent chromosome 13s. The wild-type *RB1* gene is shown as an open box and the mutant by the filled box. Heterozygous alleles at closely linked polymorphic loci are shown by the letters (not to scale); for clarity only the genotypes of chromatids that will be inherited by the subsequent retinoblastoma cell are shown. Mutation of *RB1* in case 1 is represented by a change from an open to a filled box. Deletion of *RB1* and linked markers in case 2 is represented by the thin black line. On the right are retinoblastoma cells just after mitosis containing univalent chromosome 13s. Three of the four mechanisms (deletion, non-disjunction and mitotic crossing-over) that result in loss of *RB1* function also result in loss of heterozygosity of the linked markers.

About 5–10% of hereditary retinoblastoma cases have a deletion of part or all of chromosome band 13q14. This observation identified the location of *RB1*. The isolation of *RB1* allowed these predictions to be tested. The use of RFLP markers showed LOH was common in tumour cells from hereditary cases. Moreover, the introduction of the wild-type *RB1* gene into such tumour cells suppressed their tumour characteristics, providing further experimental verification of the tumour suppressor hypothesis. LOH is now used as a key diagnostic test of whether a gene is acting as a tumour suppressor.

Although the *RB1* gene was first identified through a rare form of cancer, subsequent research has shown that its product (RB) plays a key role in regulating the cell cycle (see Box 4.3). It acts to block cells from initiating a round of mitotic cell division unless stimulated by mitogens. Other mutations that affect components in the same pathway as RB have also been found in hereditary cancers. For example, INK4a, which encodes one of the p16 family of CDK inhibitors (see Box 4.3), is mutated in familial melanoma. Many sporadic cancers have a high frequency of mutations in components of the RB pathway. For example, homozygous loss of *RB1* is found in nearly all small cell lung carcinomas.

4.7.2 Li–Fraumeni syndrome

Li–Fraumeni syndrome is a rare autosomal dominant disease characterised by predisposition to multiple forms of cancer. In women, breast cancer is the most common of these. The gene affected by Li–Fraumeni syndrome is *TP53*, which encodes p53, a key cell cycle regulator (see Box 4.3). p53 provides a checkpoint function in cells with damaged DNA. If the damage is repairable, p53 stalls the cell cycle by stimulating inhibitors of CDK4/cyclin D and CDK2/cyclin E, which are protein kinase/cyclin complexes responsible for triggering the start of the cell cycle (see Box 4.3). If the damage is severe, p53 triggers cell death by apoptosis. The removal of this function allows cells with damaged DNA to divide, increasing genome instability and increasing the chances of damage to another component of the control machinery. Inactivation of p53 is probably the most frequent mutational event in sporadic cancers.

4.7.3 Hereditary breast cancer

Breast cancer is the most common form of malignant cancer among women. Each year in the USA an estimated 182 000 women will develop breast cancer and 46 000 women will die from the disease. Estimates of the cumulative lifetime risk vary from 1 in 8 to 1 in 12. The risk of breast cancer is two- to three-fold higher in women with first-degree relatives (mother or sister) with the disease, with the higher risks associated with an earlier age of diagnosis in the relative (<40 years of age). Despite this, no specific genetic factors have been identified in the majority of these cases. However, a subset of between 5 and 10% of cases, known as **hereditary breast cancer** (HBC), are due to germline mutations. HBC shows a number of characteristics that distinguish it from the sporadic form.

- HBC clusters in families, showing a pattern of inheritance that is characteristic of an autosomal dominant trait.

- As well as breast cancer, there is an increase in other forms of cancer, particularly ovarian cancer and male breast cancer, otherwise a rare disease.

- Age of onset in HBC is much lower: the mean age of diagnosis in HBC is 47 compared to 62 in the sporadic form.

- In HBC there is a much higher incidence of bilateral cancers, i.e. tumours in both breasts.

BRCA1 and BRCA2

Mutations in two genes, BRCA1 and BRCA2, have been shown to be responsible for 60–80% of all HBC cases. A woman carrying either of these mutations has about a 40% chance of suffering from breast cancer by the age of 40, a 66% chance by age 55 and has an overall lifetime risk of 82%. BRCA1 mutations also confer a similarly high risk of ovarian cancer while BRCA2 mutations are associated with an increased risk of male breast cancer, although the risk of ovarian cancer is also elevated.

Both BRCA1 and BRCA2 were mapped by searching for co-inheritance of polymorphic genetic markers and the disease in families in which the incidence of breast cancer is high. BRCA1 is located at chromosome 17q21 and BRCA2 at chromosome 13q12–13 (this is described in more detail in Chapter 5). Both genes were cloned by positional methods, concentrating on transcripts produced in the region to which the genes were mapped. BRCA1 is composed of 22 exons distributed over more than 100 kb of genomic DNA. It encodes a protein of 1863 amino acids whose biochemical role is not yet clear, although evidence is accumulating that it is involved in maintaining genome integrity. BRCA2 is composed of 27 exons spread over 70 kb of genomic DNA; it encodes an extremely large protein of 3418 amino acids.

Cloning the BRCA1 and BRCA2 allowed mutations that predispose to breast cancer to be identified. A very large number of different mutations have now been characterised which, together with the large size of both genes, will make testing for mutations very difficult. For the most part, the mutations are frameshift and nonsense mutations that would be expected to destroy gene function and thus be recessive at a cellular level. Since the mutations in both genes are inherited as autosomal dominants, it was expected that both genes would be tumour suppressors following the paradigm established for retinoblastoma (see above). This was confirmed when it was found that LOH occurred in nearly all cases examined. Thus HBC comes about when one copy of the gene is inactivated by a germline mutation and the other is lost by a deletion in somatic cells. Furthermore, BRCA1 has been shown to function as a tumour suppressor in vitro. The proliferation of breast and ovarian cancer cell lines is inhibited by transfection with wild-type, but not mutant, BRCA1. This action is tumour specific: the proliferation of lung or colon cell lines is not inhibited by BRCA1.

It was hoped that identifying tumour suppressors specific to breast and ovarian cancers would help elucidate the molecular pathology of sporadic breast cancers. It is still not clear whether this hope will be realised because the evidence is conflicting. Expression of *BRCA*1 and *BRCA*2 is reduced in sporadic breast cancers, which might be expected of a tumour suppressor. Moreover, a significant proportion of sporadic cases show LOH at either 17q21 or 13q12–13, where *BRCA*1 and *BRCA*2 are located. However, there is a conspicuous absence of mutations in the remaining *BRCA*1 or *BRCA*2 genes, which would be predicted by the tumour suppressor model. It is possible that the mutations in the remaining copy are occurring in non-coding regions and are acting to reduce expression.

Population distributions of *BRCA1* and *BRCA2* mutations

The size of *BRCA*1 and *BRCA*2 and the large number of different mutations that can occur make it difficult to undertake population screens to measure the mutation frequencies of these genes. An indirect estimate of the frequency of *BRCA*1 alleles may be obtained from the incidence of ovarian cancer. First-degree relatives of breast cancer patients are significantly more likely to suffer from ovarian cancer compared with the general population. Since *BRCA*1 mutations are the main cause of hereditary ovarian cancer, the frequency of such mutations may be calculated from the excess risk of ovarian cancer in first degree relatives of breast cancer patients. The allele frequency arrived at by these means is 0.0006. This corresponds to a carrier frequency of 1 in 800. The carrier frequency for all HBC mutations may be as high as 1 in 300. Although such mutations account for only 5–10% of all breast cancer cases, the frequencies are very high for germline mutations and breast cancers due to *BRCA*1/2 mutations may be said to be one of the most common monogenic disorders.

Two particular HBC mutations, *BRCA*1 185delT and *BRCA*2 6174delT, have each been found to present at a frequency of about 1% in the Ashkenazi Jewish population in the USA. *BRCA*1 185delT was found to be responsible for about 28% of early-onset breast cancers in this population, but the *BRCA*2 mutation was only found to be responsible for 8% of such cancers. The proportion of cases caused by *BRCA*1 is consistent with estimates of its penetrance based on its segregation in HBC families. However, the penetrance of the *BRCA*2 mutation based on these figures is about fourfold lower than previous estimates. Another study with the same population showed that the risk of cancer in all carriers of both *BRCA*1 185delT and *BRCA*2 6174delT mutations was 56% by age 70, significantly lower than the 85% risk estimated from high-risk breast cancer families.

This discrepancy highlights a problem with the estimates of penetrance derived from high-risk families, i.e. there may be other factors contributing to the risk in these families that are not present in normal families. For example, the penetrance of a *BRCA*1 or *BRCA*2 mutation may be modified by alleles at other loci. In the breast cancer families there may be a higher frequency of unfavourable modifying alleles, leading to a higher penetrance of *BRCA*1 and *BRCA*2 mutations. In other words, there has been a bias of ascertainment that led to unusual families being selected because in these families the *BRCA1/2* alleles behaved as simple autosomal dominants. *BRCA1* and *BRCA2* may have a lower penetrance in genetic backgrounds that are more common. If this is the case, counselling a woman who carries a *BRCA1* or *BRCA2* mutation may need to take account of her family history.

4.7.4 Colorectal cancer

Colorectal cancer is the second or third most common cancer of the western world with a lifetime risk of between 5 and 6%. It is rarer in the developing world and Japan. About 90% of cases are sporadic. However, there are two well-characterised hereditary forms: **familial adenomatous polyposis** and **hereditary non-polyposis colorectal cancer**. Both forms fulfil the classic criteria of tumour suppressors.

Familial adenomatous polyposis

Familial adenomatous polyposis (FAP) affects between 1 in 8000 and 1 in 15 000 of the population. It is characterised by the presence of hundreds to thousands of benign adenomas or polyps covering the colon and rectum. These polyps are clonal in origin. Some of these polyps are larger than others and from these malignant carcinomas arise. FAP is inherited as an autosomal dominant disease. The gene involved is called **APC** (adenomatous polyposis coli). It was mapped by linkage studies to 5q21–22 and positionally cloned in 1991. *APC* is composed of 15 exons, producing a 9 kb mRNA. The cellular role of the gene is currently unknown but it is thought to be involved in cell-to-cell signalling or adhesion. Most cells in polyps are homozygous for mutations or deletions of the *APC* gene. Occasional polyps are found in normal individuals and these also bear mutations in the *APC* gene. However, loss of the *APC* gene is not sufficient for carcinomas to develop. Mutations in other genes, such as *TP53* encoding p53, *KRAS* (a member of the *ras* family) and another tumour suppressor gene known as *DCC* (deleted in colon cancer) are found in full-blown carcinomas.

Hereditary non-polyposis colorectal cancer

Hereditary non-polyposis colorectal cancer (HNPCC), also known as Lynch syndrome, is a hereditary cancer that is not associated with the formation of polyps. It is characterised by an earlier age of onset than sporadic colorectal cancer and frequently involves cancer in other organs. Estimates of its incidence vary widely because of difficulties in distinguishing it from the sporadic form. The classical form of the disease is defined by its occurrence in three members of a family in successive generations, with one affected individual diagnosed before age 50. Some estimates are as low as 1 in 10 000, others as high as 1 in 200. Estimates of what proportion of all colorectal cancers are caused by HNPCC range from 0.7 to 13%. A mean of these figures would indicate a population incidence of 1 in 400, making it one of the most common of monogenic disorders.

HNPCC is characterised by instability of multiple microsatellite loci located throughout the genome. This suggests a defect in mismatch repair of

DNA. Mismatch repair operates by excising one strand of a stretch of DNA
that contains a base-pair mismatch. The missing strand is then replaced by
the action of DNA polymerase, using the remaining strand as a template.
Mutations in genes mediating this process characteristically result in a
decrease in the stability of repeated DNA sequences.

After an HNPCC susceptibility locus was located to chromosome 2p by
linkage analysis, a search was undertaken for genes that might provide this
function. A candidate gene, human *MSH2*, was cloned that is a homologue
of the yeast *MSH2* gene and the bacterial *mutS,* both of which are known
to mediate mismatch repair. It was shown that mutations affecting *MSH2*
co-segregated with the disease in several HNPCC kindreds. Subsequent to
the identification of *MSH2*, three other genes have been cloned that are
also involved in mismatch repair and in which germline mutations co-segre-
gate with the disease in different HNPCC kindreds. These genes are known
as *MLH1, PMS1* and *PMS2*. Most cases (~80%) of HNPCC are caused by
mutations in *MLH1* or *MSH2*. Most mutations affecting these genes result
in truncated proteins or substitutions at highly conserved residues. In both
cases an inactive protein would result. Thus it seems likely that the two-hit
tumour suppressor model applies to HNPCC. Individuals with germline
mutations inactivating one copy of the gene develop tumours when the
other copy is lost due to somatic mutation.

4.7.5 Neurofibromatosis

Neurofibromatosis (NF) is characterised by so-called *café-au-lait* pigmented
skin patches and disfiguring benign tumours called neurofibroma, which are
outgrowths of **neural crests**. It exists in two forms, the more common, NF
type 1 (NF1), being one of the most widespread autosomal dominant diseases, affecting 1 in 3000 children. Mutations in the *NF1* gene predispose to a
number of tumours that originate in cells of the nervous system.

The *NF1* gene is located at chromosome 17q11 and encodes a protein
known as neurofibromin. Neurofibromin is an inhibitor of the *ras* oncogene,
acting in the same manner as another protein, GTPase-activating protein
(GAP; Figure 4.11). The tissue specificity of *NF1* mutations may reflect an
absence or lowering of GAP function in neural crest cells, making them more
dependent on neurofibromin. Tumours arising in patients with *NF1* frequently
have inactivating mutations in the remaining copy of the *NF1* gene. Therefore
the *NF1* gene behaves as a classic tumour suppressor gene, in terms of both its
cellular function as an inhibitor of an oncogene and its genetic properties.

4.7.6 Ataxia telangiectasia

Ataxia telangiectasia (AT) is another disease, like Li–Fraumeni syndrome
and retinoblastoma, where rare germline mutations help identify an important part of the normal machinery that protects against cancer. AT is a rare
autosomal recessive disease, affecting between 1 in 40 000 and 1 in 100 000
individuals worldwide. It is a complex disease, characterised by a variety of

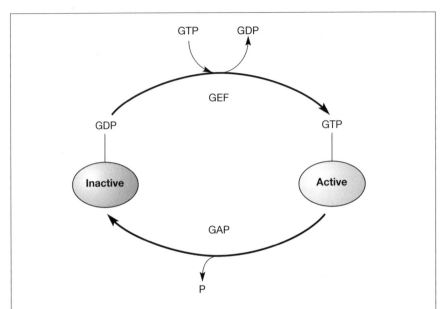

Figure 4.11 The *ras* cycle. When bound to a molecule of GTP, *ras* is inactive. Hydrolysis of GTP mediated by GTPase-activating protein (GAP) converts it to the inactive state. Exchange of GDP for GTP, mediated by guanosine exchange factor (GEF), converts it back into the active state. Oncogenic mutations lock *ras* into the active form.

apparently unconnected symptoms that first become apparent at the age of 3 and lead to death in the second or third decade:

- **ataxia** caused by cerebellar degeneration, which leads to progressive neuromotor deterioration;
- dilated blood vessels in the eye and face (telangiectasia);
- severe immune deficiency and absence of the thymus;
- acute sensitivity to ionising radiation;
- predisposition to multiple cancers;
- heterozygotes are mildly affected but exhibit predisposition to cancer and radiation sensitivity.

Genetic studies were confused by there being apparently four complementation groups, each manifesting distinct phenotypic characteristics. A single gene was located by linkage mapping, so the complementation observed is apparently an example of intragenic complementation. At a cellular level, AT cells behave as if they have a defect in the p53-mediated checkpoint pathway. The gene encodes a large protein with similarities to phosphoinositol 3-kinases, which have already been shown to be part of intracellular signalling pathways and to be involved in the control of DNA synthesis and meiosis in yeast. The checkpoint defect explains the predisposition to cancer and radiation sensitivity.

4.8 Summary

Benefits of cloning genes

- It tells us about the role of the affected protein within the cell, allowing research into new drugs to treat the disease.

- Characterisation of mutations that result in the disease allows diagnostic tests to be developed.

- Availability of the wild-type gene may lead to gene therapy.

Positional cloning

- The gene is mapped by looking for linkage between polymorphic genetic markers and the occurrence of the disease.

- Clones for the area are located in existing clone maps or isolated by chromosome walking and jumping.

- Possible genes are located by examining transcripts arising from the region.

- Co-segregation of mutations with the disease or features such as trinucleotide repeat expansions are used to decide which of the possible genes is responsible for the disease.

- The positional candidate approach examines genes known to map in the same area and which have a role consistent with the biological defect.

Cystic fibrosis

- CF is the most common autosomal recessive disease that affects Europeans.

- The gene affected is called *CFTR*. It encodes a protein that transports chloride ions across the cell membrane.

- Two-thirds of mutant alleles are the same mutation, called ΔF508.

Duchenne muscular dystrophy

- DMD is a sex-linked disease.

- The chromosomal locus is exceptionally large (2.5 Mb).

- The gene affected encodes a protein called dystrophin that links the actin cytoskeleton to the extracellular cortex via a complex called DASC.

Trinucleotide repeat expansions

- Expansions involving the trinucleotide CCG are responsible for fragile sites in chromosomes, including fragile-X syndrome. These occur in the 5' non-coding area of genes and result in their inactivation.

- Expansions involving CAG in the coding region are responsible for neurodegenerative disorders, including HD.

- MD is caused by an expansion of a CTG in the 3' UTR of a protein kinase.

- Trinucleotide repeats are unstable between generations, leading to anticipation.

Haemoglobinopathies
- These are diseases affecting haemoglobin that result in severe anaemias.

- Sickle cell disease results from a single amino acid substitution.

- Thalassaemia results in a reduction of α-globin or β-globin levels

- β^0 Thalassaemia results from mutations that prevent any β-globin expression. Milder forms of α thalassaemia can arise when deletions remove the β-globin gene but there is a compensatory persistence of foetal haemoglobin.

- There are two copies of the α-globin gene in the cluster so there are four copies in each diploid individual. The severity of α thalassaemias correlates with the number of remaining α-globin genes. The most severe is hydrops fetalis in which there are no remaining copies.

Hereditary cancers
- A complex network controls cell proliferation. Cancer results from the accumulation of somatic mutations that disrupt this network.

- Tumour suppressors are genes that normally suppress cancers; cancers result when both copies are knocked out by mutation.

- Many hereditary cancers result when one copy of a tumour suppressor gene is inactivated by a germline mutation and the other copy is lost by somatic mutation.

- Retinoblastoma is a rare form of cancer, the study of which led to the identification of a key tumour suppressor gene that is mutated in many sporadic cancers.

- About 5–10% of breast cancer is caused by mutations in two genes, *BRCA*1 and *BRCA2,* that act as tumour suppressor genes. Whether they are affected in sporadic cases is still uncertain.

- Hereditary colon cancers can also be caused by mutations in tumour suppressor genes:
  - FAP is caused by mutations in the *APC* gene.
  - HNPCC is caused by mutations in genes responsible for mismatch repair.

- NF is caused by mutations in a gene that is a negative regulator of the *ras* oncogene that is commonly mutated in sporadic cancers.

Further reading

Positional cloning

COLLINS, F.S. (1995) Positional cloning moves from perditional to traditional. *Nature Genetics,* **9**, 347–350.

POUSTKA, A. and LEHRACH, H. (1986) Jumping and linking libraries: the next generation of molecular tools in mammalian genetics. *Trends in Genetics,* **2**, 174–178

Cystic fibrosis

COLLINS, F.S. (1992) Cystic fibrosis: molecular biology and therapeutic implications. *Science,* **256**, 774–779.

KEREM, B.S., ROMMENS, J.M., BUCHANAN, J.A. *et al.* (1989) Identification of the cystic fibrosis gene: genetic analysis. *Science,* **245**, 1073–1080.

MARX, J.L. (1989) The cystic fibrosis gene is found. *Science,* **256**, 923–925.

RICH, D.P., ANDERSON, M.P., GREGORY, R.J. *et al.* (1990) Expression of cystic fibrosis transmembrane conductance regulator corrects defective chloride channel regulation in cystic fibrosis airway epithelial cells. *Nature,* **347**, 358–363.

RIORDAN, J.R., ROMMENS, J.M., KEREM, B.S. *et al.* (1989) Identification of the cystic fibrosis gene: characterisation of complementary DNA. *Science,* **245**, 1066–1073.

ROMMENS, J.M., IANUZZI, M.C, KEREM, B.S. *et al.* (1989) Identification of the cystic fibrosis gene: chromosome walking and jumping. *Science,* **245**, 1059–1065.

ZIELENSKI, J. and CHEE, L.C. (1995) Cystic fibrosis: genotypic and phenotypic variations. *Annual Review of Genetics,* **29**, 777–807.

Duchenne muscular dystrophy

AHN, A. and KUNKEL, L.M. (1993) The structural and functional diversity of dystrophin. *Nature Genetics*, **3**, 283–291.

BROWN, R.H. (1997) Dystrophin-related proteins and the muscular dystrophies. *Annual Review of Medicine*, **48**, 457–466.

KOENIG, M., HOFFMAN, E.P., BERTELSON, C.J., MONACO, A.P., FREENER, C. and KUNKEL, L.M. (1987) Complete cloning of the Duchenne muscular dystrophy (DMD) cDNA and preliminary genomic organisation of the DMD gene in normal and affected individuals. *Cell*, **50**, 509–517.

MONACO, A.P. and KUNKEL, L.M. (1987). A giant locus for the Duchenne and Becker muscular dystrophy gene. *Trends in Genetics*, **3**, 33–37.

ROBERTS, R.G. (1995) Dystrophin, its gene, and the dystrophinopathies. *Advances in Genetics*, **33**, 177–231.

Trinucleotide repeat expansions

ASHLEY, C.T. and WARREN, S.T. (1995) Trinucleotide repeat expansion and human disease. *Annual Review of Genetics*, **29**, 703–728.

Huntington's Disease Collaborative Research Group (1993) A novel gene containing a trinucleotide repeat that is expanded and unstable on Huntington's disease chromosomes. *Cell*, **72**, 971–983.

OOSTRA, B.A. and WILLEMS, P.J. (1995) A fragile gene. *Bioessays*, **17**, 941–947.

WILLEMS, P.J. (1994) Dynamic mutations hit double figures. *Nature Genetics*, **8**, 213–215.

Globin gene clusters

COLLINS, F.S. and WEISSMAN, S.M. (1984) The molecular genetics of human haemoglobin. *Progress in Nucleic Acid Research and Molecular Biology*, **31**, 314–462.

GROSVELD, F., ANTONIOU, M., BERRY, M. *et al.* (1993) The regulation of human globin gene switching. *Transactions of the Royal Society of London Series B*, **339**, 183–191.

Inherited cancers: general

BROWN, M.A. and SOLOMON, E. (1997) Studies on inherited cancers: outcomes and challenges of 25 years. *Trends in Genetics*, **13**, 202–207.

KNUDSON, A.G. (1993) Anti-oncogenes and human cancer. *Proceedings of the National Academy of Sciences USA*, **90**, 10914–10921.

SHER, C.J. (1996) Cancer cell cycles. *Science*, **274**, 1672–1677.

BOYD, J. (1995) BRCA1: more than a hereditary breast cancer gene? *Nature Genetics*, **9**, 335–336.

EASTON, D. (1997) Breast cancer genes: what are the real risks? *Nature Genetics*, **16**, 210–211.

MIKI, Y., SWENSEN, J., SHATTUCKEIDENS, D. *et al.* (1994) A strong candidate for the breast and ovarian cancer susceptibility gene *BRCA1*. *Science,* **266**, 66–71.

ODDOUX, C., STRUEWING, J.P., CLAYTON, C.M. *et al.* (1996) The carrier frequency of the BRCA2 617delT mutation among Ashkenazi Jewish individuals is approximately 1%. *Nature Genetics*, **14**, 188–190.

ROA, B.B., BOYD, A.A., .VOLCIK, K. and RICHARDS, C.S. (1996) Ashkenazi Jewish population frequencies for common mutations in *BRCA1* and *BRCA2*. *Nature Genetics*, **14**, 185–187.

TAVTIGIAN, S.V., SIMARD, J., ROMMENS, J. *et al.* (1996) The complete *BRCA2* gene and mutation in chromosome 13-linked kindreds. *Nature Genetics*, **12**, 333–337.

WOOSTER, R., BIGNELL, G., LANCASTER, J. *et al.* (1995) Identification of the breast cancer susceptibility gene *BRCA2*. *Nature*, **378**, 789–792.

WU, L.C., WANG, Z.W., TSAN, J.T. *et al.* (1996) Identification of a RING protein that can interact *in vivo* with the *BRCA1* gene product. *Nature Genetics*, **14**, 430–439.

Hereditary non-polyposis colorectal cancer

DE LA CHAPPELLE, A. and PELTOMÄKI, P. (1995) Genetics of hereditary colon cancer. *Annual Reveiw of Genetics*, **29**, 329–348.

Neurofibromatosis type 1

SHEN, M.H., HARPER, P.S. and UPADHYAYA, M. (1996) Molecular genetics of Neurofibromatosis type-1 (NF1). *Journal of Medical Genetics*, **33**, 2–17.

Ataxia telangiectasia

SAVITSKY, K., BARSHIRA, A., GILAD, S. *et al.* (1995) A single ataxia gene with a product similar to PI-3 kinase. *Science*, **268**, 1749–1753.

CHAPTER FIVE

Genetic analysis of complex disease

Key topics

- Benefits of identifying susceptibility genes
- Genetic factors in complex disease
  - Relative risk (λ)
  - Discrete and quantitative disorders
  - Polygenic control of susceptibility
  - Epistatic and additive interactions
  - Non-Mendelian inheritance and complex disease
- Identifying susceptibility loci by linkage analysis
  - Pedigree analysis
  - Allele sharing and genome scans
  - Where to set the threshold for significant linkage?
- Identifying susceptibility loci by association studies
  - Association with a biologically relevant gene
  - Linkage disequilibrium
- Type 1 diabetes as an example of a complex disease
  - Involvement of the HLA locus
  - Genome scans to detect other genes
- Alzheimer's disease

5.1 Introduction

Many of the major monogenic disorders have now been isolated by positional cloning. Monogenic disorders are often devastating in their consequences, but their population incidence is low. Diseases such asthma, migraine, diabetes, rheumatoid arthritis, multiple sclerosis, hypertension, cardiac disease, obesity and psychiatric disorders such as schizophrenia, manic depression and autism are much more common. Such diseases do not show any clear pattern of Mendelian inheritance. They do, however, cluster in families, indicating that a genetic component is likely to be operating. The aetiology of such diseases is complex, being a mix of environmental and genetic factors. For this reason they are often referred to as complex diseases.

It is possible to imagine several models explaining susceptibility to complex diseases. The simplest is that the genetic component is due to the segregation of a single major locus, modified by the environment so that it shows incomplete penetrance, and possibly complicated by the existence of phenocopies. More complicated models can invoke several genes each with a significant contribution (**oligogenes**) or many genes each making a small contribution (**polygenes**). In such models, the effect of oligogenes and polygenes on phenotype is subject to modification by environmental factors. Such models are usually referred to as multifactorial inheritance.

Why study the genetics of complex diseases?

Research in human molecular genetics is now focused on identifying the genetic factors involved in complex disease. This is not an easy task and, as we shall see, would not be feasible without the detailed genomic maps discussed in Chapter 3. However, success in this enterprise promises to prevent, or lead to early diagnosis and better treatment of, diseases that may affect us all at some stage in our lives. These benefits may be listed as follows.

1. Identification of individuals with increased susceptibility to a disease will allow lifestyle changes to lower the risk and more intensive medical surveillance to detect the first signs of its onset should it occur. Early diagnosis is nearly always a strong factor in successful treatment.
2. Establishing the underlying genetic risk factors removes one variable from the study of genotype–environment interaction. Identification of environmental risk factors thus becomes easier and more precise.
3. Many diseases with apparently similar clinical features may be heterogeneous in their molecular aetiology. This may be the reason why some patients respond well to one particular treatment and others do not. If the different types of defect that lead to the same disease can be elucidated, and suitable diagnostic tests developed, then treatments can be better fitted to the actual disease.
4. Cloning the genes contributing to the risk will lead to the characterisation of the encoded proteins. These proteins will be components of the cellular processes that are malfunctioning in the disease. This will clearly contribute to a better understanding of the normal and the pathogenic states. Such knowledge will be the starting point for the development of more rational therapies. In particular, the proteins affected may be used to establish screens for small-chemical drugs that might modify their action. For example, suppose a particular disease is shown to occur through the hyperactivity of a cell surface receptor. The receptor may be produced as a **recombinant protein**, i.e. expressed from the cloned gene in another host such as *E. coli*, yeast or mammalian tissue culture cells. This will provide a plentiful supply of the protein, which can be used in an *in vitro* screen for small drugs that reduce its activity.

5.2 Special problems of complex diseases

Efforts to identify genes contributing towards the risk of complex diseases are complicated by a number of problems that, by and large, did not arise in the study of monogenic diseases. The major problems are centred around genetic heterogeneity, reduced penetrance and phenocopy. Genetic heterogeneity refers to a situation where alleles at more than one locus can individually trigger the disease. Reduced penetrance results in an individual with a predisposing genotype not succumbing to the disease. In contrast, phenocopy refers to a situation where the disease is triggered by special environmental factors in the absence of a predisposing genotype. These considerations mean that we may expect to encounter situations where a predisposing genotype does not result in the disease and, conversely, where the disease occurs without the predisposing genotype. Clearly, this is going to make robust correlations between genotype and disease more difficult to establish. Heterogeneity also means that a gene implicated in the disease may be correctly identified in one kindred or population, but this may not be replicated in a second study.

5.3 Mapping a major locus using a parametric model

The locations of genes affected in monogenic disorders were mainly mapped by pedigree analysis, in which linkage was sought between genetic markers and the disease using LOD score analysis to combine data from multiple families (see Chapter 3). LOD score analysis tests a genetic model which states that the disease locus and a genetic marker are linked with a recombination fraction θ and which requires that parameters are specified, such as whether the trait is dominant or recessive, the degree of penetrance (which may be age related) and allele frequencies. Genetic heterogeneity may be detected by specially designed tests. The analysis is iterated, changing the parameters each time. The model with the highest LOD score (MLS) is judged the most likely to be correct. Such a procedure allows different values of θ to be tested and thus linkage of the disease trait to different genetic markers. It also allows other parameters to be varied, such as the degree of penetrance. Because this type of analysis requires parameters to be specified it is called **parametric**.

Diseases that are generally complex in their aetiology often contain subsets of cases where the disease is triggered by the action of a major locus segregating in a Mendelian fashion, perhaps with incomplete penetrance. The location of such genes could be mapped by pedigree analysis if a way can be found to restrict the analysis to those families where the disease is apparently monogenic. One common method of doing this is to redefine the criteria to select pedigrees so that only the more extreme phenotypes are used, such as early age of onset, more severe symptoms and clustering of the disease in close relatives. Examples of where this has been successful are shown in Table 5.1

Disease	Restricting criteria	Loci identified
Breast cancer	Age of onset Bilateral tumours Familial clustering	BRCA1, BRCA2
Hereditary non-polyposis colorectal cancer	Familial clustering Age of onset	MSH2, MLH1, PMS1, PMS2
Familial adenomatous polyposis	Extreme polyposis	APC
Alzheimer's disease	Age of onset	Presenilin II, Presenilin I, ApoE, APP
Premature heart disease Age of onset	Hypercholesterolaemia	Low-density lipoprotein receptor
Non-insulin-dependent diabetes	Maturity-onset diabetes of the young MODY	Galactokinase

By way of illustration, the study which mapped *BRCA*1 to 17q21 analysed 23 families with 146 cases of breast cancer. The families were chosen because they showed clustering that might indicate a familial mode of inheritance. The statistical analysis specified autosomal dominant inheritance with age- and sex-specific risks of breast cancer for susceptible and non-susceptible individuals, measured from previous large-scale population studies. The statistical analysis showed evidence for linkage to a minisatellite marker called D17S74 with an overall LOD score of 2.35. The analysis showed evidence of heterogeneity, which was resolved when age of onset was taken into account. When the analysis was restricted to seven families with a mean age of onset of <45 years the LOD score was 5.98, while the LOD score for late-onset families was negative indicating the absence of linkage.

Genetic heterogeneity for hereditary breast cancer was revealed in other studies by the existence of families that showed a clear autosomal dominant pattern of inheritance, but in which the locus responsible was not linked to 17q21. Subsequently a second locus, *BRCA*2, was mapped to 13q12–13.

5.3.1 Psychiatric disorders

Psychiatric disorders such as schizophrenia and manic depression generally show some family clustering but not Mendelian inheritance. However, like the cases discussed above, some kindreds apparently show inheritance in which the disease behaves as an autosomal dominant trait. Attempts have been made to map the genes involved. Most of the published reports of the location of these genes have not been replicated and in many cases the original report has been withdrawn or modified. For example, between 1969 and 1996 there have been 14 loci reported to be linked to manic depression. None have so far been replicated, although the most recent reports were based on non-parametric analysis using genome scans (see below), which may have avoided some of the problems of previous reports. Similarly a report of linkage of schizophrenia to a marker on chromosome 5 has not been replicated in other families.

The reasons why some studies have failed are revealing. One study followed the inheritance of manic depression in 32 families of the Amish community in the USA. Linkage was found between manic depression and a marker on chromosome 11, with an LOD score of between 4 and 5. Such an LOD score is apparently highly significant. However, two subjects who were diagnosed as normal subsequently developed symptoms of the disorder. As a result the LOD score dropped to 2, below the level of significance. The LOD score has subsequently been lowered again by further re-analysis. It was later shown that the LOD score in such linkage studies was extremely sensitive to the status of a few key individuals in the pedigree.

Another problem may be genetic heterogeneity. If different loci can be individually responsible for a disease then linkage could be correctly identified in one pedigree but may not be repeated in another where a different locus is responsible. Table 5.1 shows that genetic heterogeneity is not an insurmountable problem, because some of the examples are of diseases where more than one locus can individually result in the disease.

A final problem with psychiatric diseases may be the assignment of phenotype. Although standard psychological tests were used to define the diseases, it may be that in reality it is not possible to classify members of a pedigree as reliably as other complex diseases, such as **insulin-dependent diabetes mellitus** (IDDM). Given the importance of correct diagnosis discussed above, this could be a major problem.

5.4 Multifactorial inheritance

Multifactorial inheritance interprets the pattern of familial clustering of complex diseases by postulating that the genetic contribution is polygenic, i.e. the genetic contribution is determined by alleles at more than one locus. A simple extension of Mendelian genetics shows that in a population there will be a number of discrete classes containing different numbers of risk alleles. If the genetic risk is modified by the environment the overall susceptibility to a disease, or **liability**, is normally distributed (Figure 5.1).

Some disorders such as diabetes are discrete, i.e. they are either present or absent. Others such as obesity, serum cholesterol levels or hypertension show a continuous range in values and are known as quantitative disorders. In discrete traits it is postulated that the disease occurs when the liability passes a threshold. In quantitative diseases the measured quantity (blood pressure, body weight, etc.) will be correlated with liability. Polygenic inheritance postulates a large number of loci, each making a small contribution to

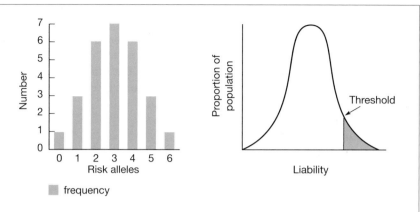

Figure 5.1 Distribution of risk alleles resulting from three contributing loci. Three **biallelic** loci each have one allele that contributes to the risk and one that does not. This results in 27 possible genotypes, ranging from where the individual with least risk has no predisposing alleles the individual with most risk who has six. The distribution of alleles follows discrete classes, but taking environmental effects into account the distribution of overall liability follows a smooth curve (right). Once a threshold of liability is passed the disease results.

the liability. It may be that one or a small number of genes make a large contribution, possibly modified by a background of polygenic inheritance. Genes making a relatively large contribution to the liability are called oligogenes.

Relative risk (λ)

Family clustering occurs because a relative of an affected individual is also likely to have inherited more predisposing alleles, as both share a proportion of their genes. The liability curve for relatives of an affected individual will therefore be shifted to the right compared with the general population. The degree of familial clustering of a discrete disease is measured by a parameter called **relative risk** or **recurrence risk** (λ) defined as:

$$\lambda_r = \frac{\text{Frequency in relatives of affected person}}{\text{Population frequency}}$$

where r denotes the degree of the relationship, i.e. first degree (sibs), second degree (uncles–nephews, grandparents–grandchildren) or third degree (cousins).

For example, the frequency of IDDM in sibs of affected individuals is 6% and the population incidence is 0.4%; λ_s for IDDM is therefore 15 (where s indicates sibs). In contrast λ_s for the monogenic disease cystic fibrosis is 500. Relative risk is a related concept to the concordance rate in identical twins and is a measure of the size of the genetic contribution towards liability.

The contribution made by a particular gene to the overall genetic contribution may also be specified in terms of λ. This locus-specific λ is defined as:

$$\frac{\text{Expected proportion of affected sib-pairs sharing zero alleles identical by descent (25\%)}}{\text{Observed proportion of affected sib-pairs sharing zero alleles identical by descent}}$$

The concept of identity by descent is discussed below. For example, a locus called *IDDM*1, which maps to the HLA locus, contributes strongly towards the liability of IDDM. The observed proportion of affected sibpairs not identical at this locus is 8%. Since the expected proportion is 25% (see below), λ_s for *IDDM*1 is calculated to be 3.1. The practical importance of locus-specific values of λ is that they are a measure of the ease of genetic mapping, being easier with traits with a high value.

Nature of interactions between genes

Alleles at different loci may combine together to generate the overall genetic contribution to the liability in contrasting ways. Their combined effect may be the sum of their individual contributions, in which case they are said to be **additive**. Alternatively the interaction may be more complex. For example, the interaction may be **multiplicative**. Consider a disease where $\lambda = 32$. Five loci with individual values of $\lambda = 2$ would produce this overall genetic contribution if they interacted in a multiplicative manner (2^5). However, it would require 16 loci with the same value of λ interacting additively to produce the same effect.

Multiplicative interaction between genes is an example of epistatic interaction. The original use of the term **epistasis** refers to an interaction between genes where one gene interferes with the expression of another. In quantitative genetics it has now become broadened to refer to a situation where the combined effect of alleles at two or more genes on a phenotype is different from that which would be predicted from their individual effects. Two important points follow from this definition. The first is that multiplicative interaction between alleles at different genes is just one example of different possible types of epistasis. The second is that when the influence on a phenotype made by a gene is assessed on its own, the result is an average of the effect of the gene taken over all the other genes with which it may or may not interact. The general methods described below to identify genes contributing to complex disease measure these average effects and do not define the interaction between particular pairs of genes unless the experiment is expressly designed to do so. Such experimental designs can be constructed, for example searches for genes contributing to IDDM have looked for the effect of genes either in the presence or absence of concordance at *IDDM*1, the locus thought to contribute most to the genetic component of liability.

Multiplicative and additive models can be distinguished by the rate at which λ declines with decreasing degrees of relatedness. The value of $\lambda_r -1$ in additive models decreases by a factor of two for each degree of relatedness (first, second, third degree, etc). In a multiplicative model the value of this term declines more steeply, for example the values of λ for schizophrenia are $\lambda_{MZ} = 52$, $\lambda_1 = 10$, $\lambda_2 = 3.2$ and $\lambda_3 = 1.8$ (where MZ, 1, 2 and 3 are monozygotic twins, first-, second- and third-degree relatives respectively). This exponential decline in the value of λ suggests a model for the genetic risk where oligogenic alleles interact epistatically.

Strategies aimed at identifying genes involved in complex diseases can be broadly divided into two groups: **linkage** and **association**. Linkage-based methods use individual families where members are affected by the disease and attempt to demonstrate linkage between the occurrence of disease and genetic markers, in essence the approach that was successfully used to map and positionally clone genes causing monogenic disorders. Association studies are based on populations. They attempt to show an association between a particular allele and susceptibility to disease.

5.5.1 Non-parametric linkage analysis

When two or more genes contribute to the genetic liability the models required to detect linkage become far more complicated. The same parameters described above for monogenic inheritance must be specified for each gene. The number of possible models is large and they can rarely be clearly distinguished by the available data. Parametric models are a powerful way of detecting linkage, but only when the models are correctly specified and individuals in the pedigree are correctly diagnosed. In general, parametric models are not useful for complex diseases where the genetic contribution is polygenic or oligogenic

If a gene is contributing towards the liability of a disease, then the region of the genome within which the gene is situated will be co-inherited from a common ancestor by affected members of a pedigree more frequently than would be expected by chance. Such a statement makes no assumptions about the way the gene is operating, nor about how many other genes may also contribute to the risk. Any method that can be used to search for such commonly inherited regions is therefore a **non-parametric** method.

Commonly inherited regions can be recognised by **genome scan**, which is now becoming the workhorse of research aimed at identifying genes that contribute towards the risk of complex disease. The genome scan follows the inheritance of polymorphic microsatellite markers in members of a pedigree. If affected members co-inherit the same allele more frequently than would be expected by chance, then the genomic region marked by that allele may contain a gene that contributes towards disease susceptibility. It is important to note that the genome scan is not, in itself, a non-parametric method. It lends itself equally well to mapping genes by parametric methods. Indeed, the linkage analysis used to map the *BRCA1* gene described above was a form of genome scan.

Genome scans are usually based on **affected sib-pairs**, i.e. two sibs in the same family who both suffer from the disease. As well as affected sib-pairs, other members of an affected pedigree can be examined (**affected pedigree member**). Markers that are evenly spaced throughout the genome are genotyped. For each autosomal locus the sibs could have zero, one or

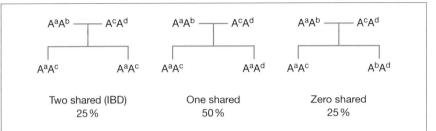

Figure 5.2 Allele sharing. The diagram shows two parents, each heterozygous at locus A (alleles: superscript a, b, c, d). There are three possible outcomes where two sibs share two, one or no alleles, which will occur with relative frequencies of 25%, 50% and 25% respectively.

two alleles in common (Figure 5.2). The expected ratio of these outcomes is 25%, 50% and 25% respectively. If the parental genotypes are known to differ at the locus, then we can specify that sibs which have both alleles in common are **identical by descent** (IBD).

If the parental genotypes are not available, we can only say that they are **identical by state** (IBS). This distinction is important because two sibs could be IBS because one or both parents were homozygous (Figure 5.3). If a number of closely linked markers can be scored, IBD can be inferred in the absence of parental genotypes because the parents would be unlikely to be homozygous for a number of different markers. If such markers are not available, IBS values must be corrected for the average population heterozygosity at the locus concerned.

In a genome scan of affected sib-pairs, a set of families containing two affected sibs is assembled. As we shall see the number of families used affects the power to detect weakly acting genes; a typical dataset would be composed of about 100 families. DNA samples are collected from the affected sibs and their parents. The microsatellite loci are genotyped using PCR with fluorescent tagged primers. Using primers tagged with different colours it is possible to analyse up to 18 loci in the same lane and up to 500 on the same gel (Figure 5.4). About 300 primers are used in the initial scan,

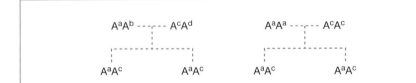

Figure 5.3 Difference between IBD and IBS. Two sibs both have the same alleles at a locus but the genotype of the parents is unknown (indicated by dotted line). If the parents were as shown on the left the sibs would be IBD at the locus. If the parents had the genotype shown on the right, IBD could not be inferred.

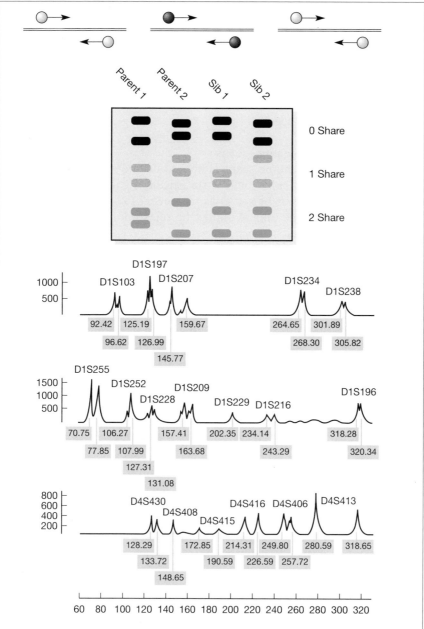

Figure 5.4 A genome scan using polymorphic microsatellite markers amplified by PCR. *Top*: schematic diagram of the analysis of three microsatellite loci in a single family. Each locus is amplified by pairs of primers that are labelled by different colour fluorescent dyes. *Bottom*: actual output from the ABI PRISM automated genotyping system analysing a single lane from the genome scan for IDDM. Each line shows the trace of a different colour fluorescent dye. Each group of peaks derive from the alleles of a single locus denoted by the D-number. The number underneath each peak records its size relative to an internal maker which identifies the allele involved. In this example 18 loci are analysed in a single lane of the gel.

which will cover the genome at 10 cM intervals. At each locus there will be two alleles in each subject (except for sex-linked loci in boys) distinguished by the length of the amplified PCR products. We can therefore specify for each locus the parental genotype and the genotype of both sibs. From this it is possible to determine for how many alleles the sibs are IBD. Using the combined data from all the families the IBD frequencies at each locus can be determined and tested to see whether they deviate from the value expected from random segregation.

Statistical analysis of genome scans

The key question in genome scans is how to decide whether an excess of allele sharing at a particular locus is statistically significant. How can the levels of statistical rigour be set so as to prevent false-positive identifications, while at the same time avoiding the rejection of weak but genuine effects? Two types of test can be applied.

1. The maximum LOD score (MLS) is related to the LOD score used for testing linkage. It measures the logarithm of the likelihood ratio that the proportion of allele sharing has the observed value, compared with the hypothesis that there is no excess sharing.
2. The P value measures the probability that a deviation as high as that observed could occur by chance under independent assortment.

The traditional test of significance in any statistical procedure is that the observed result will occur by chance less than 1 time in 20. However, in a genome scan hundreds of separate loci are tested. If we used the traditional threshold of significance we could expect many false positives. The straightforward way of dealing with this problem would be to recalculate the threshold of significance so that the chance of seeing a deviation as large as that observed would only occur in less than 1 in 20 genome scans. On this basis the values of MLS and P would be 3.6 and 2×10^{-5} respectively. Now here's the rub: it is unlikely that such high values would ever be achieved in real life with genes with low λ values.

The power of a genome scan is the probability that linkage between a gene and a marker will be detected and is measured as a percentage, e.g. a 95% probability or a 60% probability, etc. The power to detect linkage for any one locus is dependent on the λ_r value of the locus, where r specifies the relationship of the affected members (usually sibs). The larger the value of λ, the greater the chance of detecting linkage.

An example from theoretical calculations illustrates how this affects the outcome of genome scans. To detect a locus where $\lambda_s = 1.6$ with a 90% probability would require 300 families with markers spaced 1 cM apart with 100% heterozygosity. Microsatellite markers are on average only 70% heterozygous and initial scans use markers 10 cM apart. However, **multipoint mapping** (using more than one marker at a time) increases the power to detect linkage and allows the use of less tightly linked markers.

This calculation illustrates that the ideal experimental design would involve large numbers of families. However, it is not easy to find large numbers to take part in the study. One compromise is to accept a lower level of stringency, e.g. MLS = 1, in initial studies but to realise that this only suggests linkage which falls short of significance. It is then necessary to follow up with further studies, using more closely spaced markers with the same set of families and then with a different set of families. Another possibility is to complement linkage studies with a different strategy, such as **linkage disequilibrium** (see below). Even accepting the most rigorous statistical standards, 1 in 20 reports of linkage may be false.

One last point to note in connection with genome scans is that they are only possible because of the construction of detailed genetic maps with densely spaced markers, such as those from the Genethon map discussed in Chapter 3. Like the construction of the maps, genome scans are also big science. A single genome scan requires 120 000 separate assays. Moreover, assembling and medically assessing groups of perhaps 100 families requires large clinical resources.

5.5.2 Association studies

In contrast to linkage-based studies, which work with family pedigrees, association studies are based on populations. An attempt is made to find an allele associated with the disease. This can come about for two fundamentally different reasons. The allele may actually affect a biologically relevant locus or the association may be due to linkage disequilibrium.

Examples of the contribution of biologically relevant loci recognised by association are:

- autoimmune diseases associated with alleles at the HLA locus:
  - IDDM;
  - multiple sclerosis;
  - ankylosing spondylitis, over 90% of cases have the allele HLA-B27;
  - rheumatoid arthritis;
- angiotensin-converting enzyme and heart disease;
- low-density lipoprotein receptor and heart disease;
- insulin locus and IDDM.

Linkage disequilibrium refers to a non-random association in a population of alleles at two closely linked loci, i.e. a tendency of alleles at one locus to be found in combination with certain alleles at a second, closely linked locus. It may occur when most cases of a disease in a population are caused by a mutation in a common ancestor. Alleles closely linked to the site of the mutation will be commonly inherited by all individuals who have inherited the mutation (Figure 5.5).

In time the disequilibrium will be eroded by recombination. The time it takes to do so will depend on the degree of linkage. If two alleles, 1 Mb apart, are in linkage disequilibrium then it will take about 70 generations

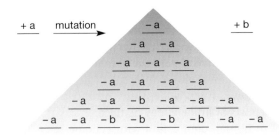

Figure 5.5 Linkage disequilibrium. Two alleles *a* and *b* exist at a polymorphic locus which is located near to a gene that can mutate to give rise to susceptibility to a disease. The wild-type allele is designated + and the predisposing allele is −. In a population a mutation occurs on a chromosome that carries the *a* allele at the linked locus. Initially, all chromosomes that descend from the chromosome with this founding mutation will be of the genotype −*a* and there are no −*b* chromosomes. The alleles at the two loci are therefore said to be in disequilibrium. In time, recombination will erode the disequilibrium and the frequencies of −*a* and −*b* chromosomes will equalise. The degree of disequilibrium is shown in the diagram by the blue shading.

(~2000 years) for the degree of disequilibrium to decay by 50%. More distantly located alleles will decay to a random configuration more quickly than closely linked alleles. Thus, there will be a gradient of disequilibrium, with the highest values closest to the gene. Thus linkage disequilibrium is essentially a linkage-based method that makes use of all the meioses in a population that have happened since the original mutation and can be used to map the location of a gene. It has proved a powerful tool for fine-structure mapping and in many cases can narrow the location of a gene to a smaller interval than linkage mapping. For example, the location of a gene responsible for a rare monogenic disorder called **diastrophic dysplasia** was mapped to within 60 kb using linkage disequilibrium in a Finnish population. Previous linkage studies had only mapped it to 1.5 cM (~1000 kb). Linkage disequilibrium was an essential tool for locating the HD gene. As well as being used to map a gene, linkage disequilibrium works in the converse sense. Disequilibrium between disease-causing alleles and closely linked alleles is evidence of a founder mutation, as opposed to repeated independent mutation events.

Linkage disequilibrium studies are heavily influenced by population structure. The effect will be greatest in small homogeneous populations, such as in Finland, and least in large heterogeneous populations, such as in the USA, because there is more likelihood of a single founder mutation in the former type of population. Population heterogeneity may lead to false-positive associations because the frequency of a disease may be different in different ethnic groups. If this is the case, any allele more common in the more affected ethnic group will show a positive association with the disease even though most loci will be biologically unconnected. For this reason careful controls are necessary, such as unaffected members of the same family. **Transmission disequilibrium tests** (TDTs) are used to check the results of an association study; they confirm whether a parent heterozygous for an associated and a non-associated allele transmits the associated allele more often to affected offspring.

Association and linkage studies can sometimes give apparently contradictory results. Alleles at a marker locus can be linked to a disease but not associated. This is because there may be several different predisposing alleles in a population that are carried by chromosomes with different haplotypes. No one haplotype will therefore be associated with the disease. Conversely, a marker may be associated with a locus but not apparently linked. This occurs when the frequency of the associated allele is high but its effect is small.

5.6 Analysis of genetic factors predisposing to IDDM

Diabetes mellitus is a disease characterised by the inability of body cells to take up glucose, leading to high levels of glucose in blood and urine but intracellular starvation. In order to gain energy, cells break down fats. The end-product of fat breakdown is acetyl-CoA, which is metabolised to acetoacetone, β-hydroxybutyrate and acetone, collectively known as **ketone bodies**. These accumulate in the bloodstream (**ketonaemia**) and are excreted in the urine (**ketonuria**). The acetone contaminates the breath of diabetic patients with a characteristic odour. The build-up of ketone bodies is known as **ketosis**. The uptake of glucose into cells is promoted by the hormone insulin produced in the islets of Langerhans in the pancreas.

There are two different diseases that result in elevated blood glucose levels.

1. **Type 1 diabetes** or IDDM. This disease is characterised by lack of insulin and can usually be treated by insulin injections. In most cases the lack of insulin is caused by the autoimmune destruction of the insulin-producing β cells in the islets of Langerhans. Typically the disease shows rapid juvenile onset, the median age of diagnosis being 12 years. Although the immediate symptoms can be treated with insulin, in the long term IDDM sufferers may face complications that lead to blindness, kidney failure, amputation of lower limbs and heart attacks. It affects about 4 in 1000 Caucasian children of European descent. There are striking ethnic differences in its frequency, the condition being rare in Mongoloid and Negroid populations.

2. **Type 2 diabetes** or **non-insulin-dependent diabetes mellitus** (NIDDM). As its name implies, this disease is not caused by a lack of insulin, but is characterised by high blood and urine glucose levels that are resistant to insulin. Typically it affects people over 40 and progresses more slowly than IDDM. There are clearly predisposing genetic factors that are apparently unmasked by environmental factors. Insulin resistance is common in obese people and may result from prolonged exposure to high glucose levels resulting from sugar-rich diets or there may be genetic factors that predispose both to diabetes and obesity. NIDDM accounts for about 85% of diabetes cases; rates are much higher than

IDDM in developed societies, probably reflecting the effects of affluence on diet. A subtype of this disease is **maturity-onset diabetes of the young** (MODY), which shows earlier onset than normal diabetes, affecting teenage children and young adults. It often shows a monogenic segregation. Half of MODY cases have been shown to be associated with mutations in the galactokinase gene.

In children of European–Caucasian descent, IDDM is the second most common cause of chronic ill health after asthma. In the USA, there are more than 14 million cases of diabetes. According to the National Diabetes Research Commission the annual consequences of diabetes in the USA include:

- 15 000–39 000 new cases of blindness

- 13 000 cases of end-stage renal disease

- 54 000 amputations, mostly of lower limbs

- 162 000 deaths from heart attacks, strokes, etc.

Both types of diabetes show strong evidence of a genetic component, although there is also a strong environmental component, as demonstrated by the continued rise in its incidence. As an example of genetic research into the causes of a common serious disease we consider below how the genetic factors predisposing to IDDM can be identified. This research is perhaps the most advanced of its type and has become the model for the analysis of other complex diseases. The population incidence of diabetes in the western world is 0.4%, while the frequency of affected individuals in sibs is 6%; therefore $\lambda_s = 15$ (see above). The concordance in identical twins is 36%. Clearly, genetic factors influence the risk of disease. Note, however, that the twin concordance is very much lower than 100%, emphasising the importance of environmental factors.

Involvement of alleles at the HLA locus

The first step towards identifying loci involved in susceptibility to IDDM was to look at candidate genes suggested by the pathophysiology of the disease. As IDDM is an **autoimmune** disease, a good place to start are the genes responsible for the distinction between self and non-self. In humans these are genes of the **human leucocyte antigens** (HLA) region on chromosome 6. They encode the major histocompatibility complex (MHC) antigens that mark the body's cells as self and protect them from immune attack (Box 5.1)

Groups of diabetic patients and control groups matched for age and ethnicity were examined to see if alleles at any HLA loci were significantly more common in affected individuals, which would suggest that those alleles lead to a susceptibility to diabetes. Conversely, one can ask whether any alleles are significantly more common in healthy controls, suggesting that those alleles exert a protective effect.

BOX 5.1: THE HLA LOCUS

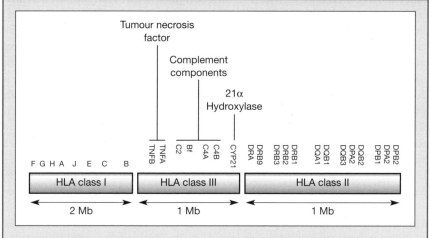

When organs are transplanted from one human to another the transplanted organ is rejected because the host recognises the organ as foreign. Rejection is controlled by the major histocompatibility antigens displayed on the cell surface. Many of these are encoded by a cluster of genes called the HLA locus (human leucocyte antigens). The HLA locus occupies 4 Mb of DNA on chromosome 6 and contains over 100 genes. The region is divided into three classes: class I, class II and class III. Organs are only accepted as self if they are identical to the host at class I and class II genes, which are highly polymorphic. The proteins encoded by these genes are displayed on the cell surface of all cells. During the neonatal period, they induce tolerance in cytotoxic T cells in the thymus. The extracellular domains of both class I and class II proteins are part of the immunoglobulin superfamily. Class III genes encode proteins and peptides with diverse functions in the immune system, for example the production of cytokines such as tumour necrosis factor and complement fixation.

As well as marking self and non-self, class I and class II molecules are responsible for this presentation of foreign **antigens** to T-helper cells. Class I molecules present intracellular antigens from all cells to T cells, which allows the recognition of intracellular infective agents such as viruses, resulting in the stimulation of cytotoxic T cells that kill the infected cells. Class II molecules present extracellular antigens on dendritic cells, B cells and macrophages, so triggering the humoral or antibody response. Class II molecules are heterodimers of α and β chains, encoded by pairs of closely linked genes. For example, DQA1 and DQB1 encode DQα1 and DQβ1, which form a heterodimer. The two chains come together so that a groove is formed between them in which the foreign antigen is located and presented to receptors on the T-helper cell.

The first results indicated that *DR3* and *DR4* alleles of a class II gene called *DRB1* (see Box 5.1) lead to susceptibility, while *DR2* alleles are protective. Further analysis showed that two more loci, *DQB1* and *DQA1*, closely linked to *DRB1* were the loci mainly responsible. Initially the involvement of *DQB1* and *DQA1* was masked because in Caucasian populations the *DR3* and *DR4* alleles were in strong linkage disequilibrium with alleles at *DQA1* and *DQB1*; they were inherited *en bloc*, leading to protection or susceptibility. Protective alleles are dominant to susceptibility alleles; a single dose of a protective *DRB1* allele confers protection. The relative frequencies of these alleles in different ethnic groups adequately accounts for the racial differences in the occurrence of IDDM. This locus is now referred to as *IDDM1*.

The aetiological mutations at *IDDM1* have now been identified. Aspartic acid at residue 57 in DQβ1 (the protein encoded by *DQB1*) is protective, while any other amino acid leads to sensitivity. This may affect the way in which antigens are presented to T-helper cells. The **autoantigen** (the protein or part of a protein recognised by the body's immune system) involved has not so far been identified, but we are clearly close to understanding one part of the chain of events that leads to diabetes, illustrating the value of this type of research.

Insulin locus

A second locus biologically relevant to diabetes is the insulin locus (*INS*) on chromosome 11. When it was cloned in 1981 it was examined to see if any alleles were associated with diabetes. Restriction mapping revealed a length polymorphism that was caused by a minisatellite or VNTR 5' to the insulin coding region. Although the number of repeats was highly polymorphic, they could be broadly divided into three classes:

1. class I is composed of 26–63 repeat units;
2. class II is composed of 80 repeat units and is rare in Caucasian populations;
3. class III is composed of 141–209 repeat units.

Class I homozygotes showed a positive association with *IDDM1* in Danish, British and US populations. However, it was puzzling that there appeared to be no linkage between the *INS* VNTR and diabetes in family pedigrees. This paradox was resolved by the application of the TDT. A pedigree that failed to show linkage between *IDDM1* and *INS* was re-examined, restricting the analysis to those families where one of the parents was a class I/class III heterozygote. When this was done, significant evidence for linkage was revealed. The original linkage was missed because of the high frequency but low penetrance of the class I allele.

Is the class I VNTR the aetiological mutation or is it in disequilibrium with the real culprit? The chromosomal region contains a number of other polymorphic sites. This provides a good example of the power of linkage disequilibrium for high-resolution mapping. A large group of families (446) with at least two diabetic offspring were examined using TDT to compare the association of each of the linked alleles with diabetes (Figure 5.6). The linkage disequilibrium curve across 40 kb rises to a peak of 66% transmission at the *INS* VNTR.

Such a result is consistent with the VNTR being the aetiological mutation; however, it is also possible that the aetiological mutation is located very close to the VNTR. Ten other polymorphisms are located within a 4.1-kb region surrounding the *INS* locus; could one of these be responsible? Another type of analysis, **cross-match haplotype analysis**, proved that the VNTR is indeed the cause of the increased susceptibility to diabetes (Figure 5.7). Because this locus has been demonstrated to influence the risk of IDDM it is known as *IDDM2*.

How does the VNTR act to influence the occurrence of diabetes? The effects of the VNTR on insulin transcription are not clear, but it seems likely that it will affect transcription in some way. The repeated sequence contains a binding site for a protein called pur-1 that can induce insulin expression. The same protein is also known as MAZ, a transcription termination factor that can reduce transcriptional read-through and thus possible interference from the adjacent *TH* locus.

Genome scans
IDDM1 alone contributes about 35% of the genetic component of susceptibility to IDDM and is almost certainly the largest single genetic factor,

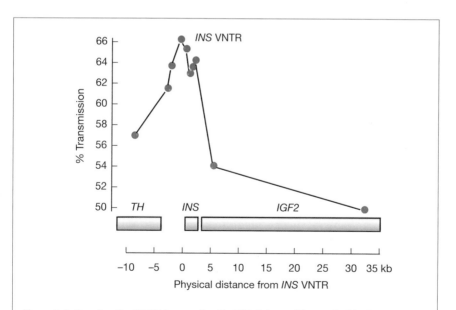

Figure 5.6 Mapping the *IDDM2* locus using the TDT. Polymorphisms at loci in the *INS* region were genotyped in 446 families with two diabetic offspring. Alleles associated with IDDM were transmitted from heterozygous parents significantly more frequently than 50%, i.e. the alleles and IDDM were in linkage disequilibrium. The degree of disequilibrium rises to a peak at, or very close to, the *INS* VNTR.

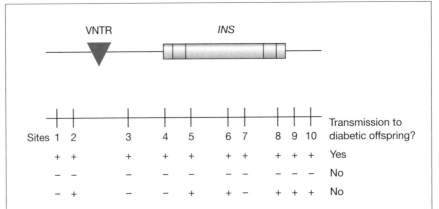

Figure 5.7 Cross-match haplotype analysis to exclude polymorphisms closely linked to the VNTR as the primary determinants of *IDDM2*. In a group of 425 diabetic families most chromosomes were of one of two haplotypes, being either +++++++++ or − − − − − − − − −. The +++++++++ chromosome was preferentially transmitted to diabetic offspring from heterozygous parents (hence the alleles it carried were designated '+'). Rare chromosomes show a mixture of genotypes, allowing certain sites to be excluded. In this example, parents who were +++++++++/−+−−++−+++ heterozygotes preferentially transmitted the +++++++++ chromosome to diabetic offspring, allowing sites 2, 5, 6, 8, 9, 10 to be excluded because the '+' allele could occur on both susceptible and protective haplotypes. Further examples of this type proved that the primary determinant was the VNTR.

although the combined effect of other genetic factors is larger. *IDDM2* accounts for about another 10%, so that together *IDDM1* and *IDDM2* account for about 50% of the risk. An idea of the number of other genes that could play a part is provided by studies using the **non-obese diabetic (NOD) mouse**, which develops a disease closely similar to IDDM with similar frequencies. The NOD mouse model can be used for detailed genetic analysis not possible in humans. Such studies confirm the role of the MHC locus (the mouse equivalent of the human HLA locus) and in addition suggest that at least 14 other loci are involved. How can we identify these loci in humans?

Genome scan based on allele sharing in affected sib-pairs has now been used by a number of different laboratories. Table 5.2 shows an example of the results obtained from one study. This has led to the identification of 12 different loci (Table 5.3). The identification of *IDDM1*, *IDDM2*, *IDDM4*, *IDDM5* and *IDDM7* have been independently replicated so are likely to be real. However, some studies have failed to replicate the positive findings of others. As we saw above, this is something that we should expect. Interestingly, one study successfully identified *IDDM2* in one dataset but failed to repeat the finding with another group of families. Since there is ample evidence confirming the role of *IDDM2*, we may regard this as an example of the difficulty in replicating genuine linkage.

If the individual values of λ for each locus are multiplied together, the combined λ is close to the overall genetic risk, suggesting that the interactions are largely multiplicative.

Table 5.2 A genome scan for IDDM. A set of 96 families from the UK where two sibs suffered from IDDM were genotyped at 290 marker loci (see Figure 5.4). Because the markers were highly polymorphic, it was possible to distinguish between maternally and paternally derived chromosomes, and so determine for each chromosome whether the sibs were IBD. In this situation, a significant deviation from a 1:1 ratio suggests linkage between the marker and IDDM. The value of MLS for significance was set at 2.3 (see pages 140–1 for a discussion of this controversial topic). A total of 20 chromosome regions showed some evidence of linkage, although only some are expected to be genuine. The table below shows data for four loci where the evidence of linkage is strong. Two of these correspond to *IDDM*1 (the HLA locus) and *IDDM*2 (the insulin locus) that had previously been implicated by other studies (see text). The effect of the *IDDM*1 genotype on other loci was analysed. This increased the power to detect linkage not apparent in the total dataset (FGF3) or increased the level of significance (ESR). Biologically, this suggests epistatic interactions between *IDDM*1 and each of these other two loci (designated *IDDM*4 and *IDDM*5). Two additional sets of families were used to analyse further cases where there was some positive evidence for linkage. When the data for all three family sets were combined, the MLS for FGF3 was 2.8 in sib-pairs discordant at *IDDM*1. Moreover, markers closely linked to FGF3 also linkage to IDDM. Taking these into account, the MLS at FGF3 increased to 3.4. Interestingly, in these other two family sets, there was no evidence for linkage at *IDDM*2 even through the first dataset and other evidence (see text) had convincingly implicated this locus.

Chromosome	Marker	Locus	IBD		MLS	Data used
			1 share	0 share		
6p	TNFα	IDDM1	97	35	7.3	All
6q	ESR	IDDM5	95	59	1.8	All
			61	30	2.5	Sibs sharing two alleles at IDDM1
11q	INS	IDDM2	107	67	2.1	ALl
11p	FGF3	IDDM4	71	60	–	All
			38	21	1.8	Sibs sharing zero/one alleles at IDDM1

Table 5.3 Loci that may contribute to susceptibility to IDDM in UK families. Note that the λ_s value for IDDM1 given here varies from that previously given (3.1) (see section 5.4).

Locus	Chromosome	λ_s
IDDM1	6p21	2.6
IDDM2	11p15	1.29
IDDM3	15q	–
IDDM4	11q13	1.07
IDDM5	6q25	1.16
IDDM6	18q	1.10
IDDM7	2q31	1.13
IDDM8	6q27	1.42
IDDM9	3q21–25	1.26
IDDM10	10 cen	1.45
DXS1068	X	1.21
GCK1	7p	–

5.7 Genetic susceptibility to Alzheimer's disease

The analysis of the aetiology of Alzheimer's disease is a good illustration of how the various techniques described in this chapter have been used in concert to study a major complex disease and reveals the difficulties that are encountered in the process.

Alzheimer's disease is the most common form of senile dementia. In the USA, the population incidence is 0.1% at age 60 rising to 2.2% after age 80. Less common early-onset cases can occur before age 60. It is characterised pathologically by the presence of neurofibrillary tangles in the cerebral cortex, together with amyloid plaques – complex proteinaceous deposits found in the brain. Neurone degeneration and death occur in the hippocampus and parts of the cerebral cortex, regions that function in memory and learning.

A number of environmental risk factors have been proposed, including head trauma, exposure to heavy metals and a family history of Down's syndrome. Genetically the disease appears to cluster in families, although the genetic epidemiology is complicated by a number of factors.

- The disease symptoms are mimicked by a number of other forms of dementia.

- Because of its late age of onset, first-degree relatives of affected individuals are either too young to show symptoms (children) or have died long ago so that information is only available by anecdotal family evidence or medical records that may not have reliably distinguished between Alzheimer's disease and other forms of dementia.

- Identical twin studies are complicated by competing causes of death in the twin who is not the proband.

- Extended families apparently showing an autosomal dominant pattern of inheritance are probably distorted by bias of ascertainment, i.e. they are only noticed and reported because they are unusual and they may not reflect a true picture of the normal pattern of inheritance.

Despite these difficulties it is clear that there is a strong genetic component to the liability of Alzheimer's disease. Estimates of the risk to first-degree relatives vary from 24–50%, while the twin concordance rate is estimated at 40–50%. Statistical analysis of collective results from a number of extended pedigrees further supports the involvement of a genetic component.

Analysis of the rare early forms of Alzheimer's disease has led to the identification of three loci that contribute to a small minority of the total. An association between the occurrence of Down's syndrome and Alzheimer's disease in families led to a report of linkage to a locus on chromosome 21. This turned out to be incorrect, but subsequently it was shown that a very small proportion of early-onset cases were associated with an allele of the *APP* gene on chromosome 21, which encodes the **amyloid precursor protein** found in senile plaques. Two more loci were shown to be responsible for the disease in a small proportion of pedigrees showing autosomal dominant inheritance of the early-onset form. Both genes encode related proteins, presenilin I and presenilin II. The proteins are homologous to the *Caenorhabditis elegans* sel-12 protein, which is thought be a receptor involved in intercellular signalling. The gene encoding presenilin I is located on chromosome 14 while that for presenilin II is located on chromosome 1. The latter is responsible for a cluster of Alzheimer families in the Volga basin region of Germany.

The use of non-parametric genome-scanning methods detected linkage between *APOE* and the much more common late form of Alzheimer's disease. *APOE* encodes **apolipoprotein E**, a plasma protein involved in lipid transport and metabolism. Interestingly, it is also a risk factor in cardiovascular disease. ApoE protein has been detected in senile plaques and neurofibrillary tangles and also binds amyloid protein *in vitro*. Thus both biological and genetic data lead to the *APOE* locus as a major risk factor for Alzheimer's disease. There are three major alleles at this locus: *APOE*E2* (approximately 6%), *APOE*E3* (approximately 78%) and *APOE*E4* (approximately 16%). The alleles differ from each other according to the identity of amino acid residues at two positions in the polypeptide chain. Association studies showed that *APOE*E4* increases risk in a dose-dependent fashion, while a single dose of *APOE*E2* is protective. *APOE* is thought to be responsible for the genetic components of the liability in about 40–50% of Alzheimer's disease cases. Its identification allows further dissection of the environmental risk factors. For example, head trauma has been confirmed to be a risk factor, but only in patients that carry at least one *APOE*E4* allele. This observation illustrates how defining the genetic components of liability facilitates the analysis of environmental risk factors in complex disease.

5.8 Summary

- Susceptibility to many common diseases is influenced by genetic factors.

- Identifying these factors would help to recognise those at risk, define environmental risk factors and lead to novel and more rational methods of treatment.

- Lack of concordance between the genotype and phenotype poses special problems in the study of complex disease. Some individuals may have the predisposing phenotype but not the disease, while others may have the disease but not the phenotype.

- In some complex diseases a subset of cases may be caused by the segregation of a single major locus. These can be mapped by pedigree analysis using parametric models. A successful example of this is the mapping of *BRCA1*.

- The multifactorial model for complex disease postulates that liability is determined by the interaction of more than one gene (oligogenes or polygenes) modified by environmental risks. For discrete diseases such as IDDM, the disease is triggered once liability passes a certain threshold.

- The size of the genetic component is defined by the parameter λ, which compares the population frequency to the frequency in relatives of an affected person. Values of λ for individual loci indicate the ease with which the locus can be mapped.

- The effect of polygenes may be the sum of the separate risks (additive) or there may be epistatic interactions between the genes. One example of such an interaction is where the outcome is the product of the individual values of λ for each locus.

- One way of identifying genes contributing towards liability is to search for regions of the genome that are IBD in affected members of a pedigree. This is done using a genome scan with highly polymorphic markers to detect excess allele sharing in affected sib-pairs.

- Susceptibility genes may also be detected by association of an allele with the disease in a population. This may come about for two reasons. The associated allele may affect a biologically relevant gene or the allele may be in linkage disequilibrium with the aetiological mutation.

- Linkage disequilibrium is the non-random association in a population of alleles at closely linked loci. Analysis based on this phenomenon makes use of all the meioses in a population that have happened since a founder mutation. It is therefore a powerful method of fine mapping.

- The analysis of genetic factors predisposing to IDDM (type 1 diabetes) illustrates how these methods can be used. Association studies have identified different alleles at the HLA and insulin loci that either protect or lead to susceptibility to IDDM. Transmission disequilibrium analysis led to the identification of the aetiological mutations.

- Genome scans have been used to confirm the involvement of the HLA and insulin loci and have led to the identification of a further eight loci.

Further reading

General

GHOSH, S. and COLLINS, F.S. (1996) The geneticists approach to complex disease. *Annual Review of Medicine*, **47**, 335–353.

LANDER, E.S. and SCHORK, N.J. (1994) The genetic dissection of complex traits. *Science*, **265**, 2037–2048.

WEEKS, D.E. and LATHROP, M.G. (1995) Polygenic disease: methods for mapping complex disease traits. *Trends in Genetics*, **11**, 513–519.

Mapping *BRCA1*

FORD, D. and EASTON,. D.F. (1995) The genetics of human breast and ovarian cancer. *British Journal of Cancer*, **72**, 805–812.

HALL, J.M., MING, M.K., NEWMAN, B., ANDERSON L.A., HUEY, B. and KING, M.C. (1990) Linkage of early onset breast cancer to chromosome 17q21. *Science*, **250**, 1684–1689.

WOOSTER, R., NEWHAUSEN, S.L., MANGION, J. *et al.* (1994) Localisation of a breast cancer susceptibility gene, BRCA2, to chromosome 13q12–13. *Science*, **265**, 2088–2090.

Epistasis

FRANKEL, W.N. and SCHORK, N.J. (1996) Who's afraid of epistasis? *Nature Genetics*, **14**, 371–373.

Psychiatric disorders

ASHALL, F. (1994) Genes for normal and diseased mental states. *Trends in Genetics*, **10**, 37–39.

MARSHALL, E. (1994) Highs and lows on the research roller coaster. *Science*, **264**, 1693–1695.

PERICAK-VANCE, M.A. and HAINES, J.L. (1995) Genetic susceptibility to Alzheimer's disease. *Trends in Genetics*, **12**, 504–507.

PLOMIN, R., OWEN, M.J. and MCGUFFIN. P. (1994) The genetic basis of complex human behaviours. *Science*, **264**, 1733–1739.

RISCH, N. and BOSTEIN, D. (1996) A manic depressive history. *Nature Genetics*, **12**, 351–353.

HLA locus and linkage disequilibrium

TOMLINSON, I.P.M. and BODMER, W.F. (1995) The HLA system and analysis of multifactorial disease. *Trends in Genetics*, **12**, 493–497.

The theoretical concept of linkage disequilibrium was first elaborated from the study of tightly linked alleles at the HLA locus. As well as reviewing the HLA locus, this paper also considers the theoretical background to linkage disequilibrium.

Statistics

LANDER, E. and KRUGLYAK, L. (1995) Genetic dissection of complex traits: guidelines for interpreting and reporting linkage results. *Nature Genetics*, **11**, 241–247.

Mapping genes predisposing to IDDM

BENNETT, S.T. and TODD, J.A. (1996) Human type 1 diabetes and the insulin gene: principles of mapping polygenes. *Annual Review of Genetics*, **30**, 343–370.

CORDELL, H.J. and TODD, J.A. (1995) Multifactorial inheritance in type 1 diabetes. *Trends in Genetics*, **12**, 499–503.

DAVIES, J.L., KAWAGUCHI, Y., BENNETT, S.T. *et al.* (1994) A genome-wide search for human type 1 diabetes susceptibility genes. *Nature*, **371**, 130–136.

CHAPTER SIX

Gene therapy

Key topics

- Somatic and germline gene therapy
- Gene replacement and gene addition
- *In vivo, ex vivo,* and *in vitro* gene therapy
- Transgenic animal models
- Vehicles for gene transfer
  - ○ Retrovirus
  - ○ Adenovirus and adeno-associated virus
  - ○ Liposomes and lipoplexes
  - ○ Naked DNA
- Gene therapy for cystic fibrosis
- Gene therapy for Duchenne muscular dystrophy
- Gene therapy for non-heritable disorders
  - ○ Cancer
  - ○ DNA vaccines

6.1 Introduction

In theory, genetic diseases may be treated by the introduction of the wild-type gene into the cells affected by a mutation. Such an approach is called gene therapy. Following the successful isolation of the genes affected in many common monogenic disorders, there is considerable expectation that gene therapy will, for the first time, offer the prospect of a cure for genetic diseases. Moreover, the introduction of DNA molecules and oligonucleotides into cells can be used to provide novel treatments for many kinds of non-hereditary diseases such as cancer. In this chapter various approaches to gene therapy are described and progress that has been made in their application reviewed. It will be seen that experiments *in vitro* and with animal models demonstrate that, in principle, gene therapy is possible. Nevertheless, the results of the first clinical trials demonstrate that successful application to human patients for the treatment of hereditary diseases is still a long way off; however, gene therapy to treat other types of diseases such as cancer is giving more promising results.

6.2 Types of gene therapy

Somatic and germline gene therapy

There are two strategies for correcting an inherited disease by gene therapy (Figure 6.1). In **somatic gene therapy** the genetic defect is corrected only in the somatic cells of a person affected by the disease. In **germline gene therapy** a genetic modification is made to a gamete, fertilised egg or embryo before the germline has split off from the cells that will make the rest of the body. The crucial difference between these two strategies is that in the first, any genetic changes are restricted to the lifetime of the person treated, while in germline therapy any change is passed on to subsequent generations. Somatic gene therapy poses few ethical problems, but germline gene therapy represents a fundamentally new type of human activity whose consequences need to be thought through carefully before any experiments are attempted. Because of this, such experiments are currently prohibited in most countries.

In vivo and *ex vivo* gene therapy

One strategy to introduce the **transgene** (the exogenous gene) is to isolate cells so that they can be manipulated *in vitro*. This would allow the cells that have received the transgene to be selected and perhaps cultured *in vitro*. These cell clones would then be reintroduced into the patient's body. Such a strategy is called *ex vivo* **gene therapy** (Figure 6.1). Alternatively the transgene could be directly introduced into the cells affected by the disease; this is called *in vivo* **gene therapy**.

Ex vivo gene therapy is suitable for genetic defects that affect blood cells. All blood cells are derived from pluripotent stem cells in the bone marrow. They can be isolated, cultured *in vitro* and successfully reintroduced *in vivo*. As described below, *ex vivo* therapy was used for the treatment of adenosine deaminase (ADA) deficiency, the only gene therapy experiment that has so far resulted in any sign of clinical improvement. Clinical trials using *ex vivo* approaches are currently underway for Gaucher's disease, Fanconi's anaemia, Hurler's syndrome and chronic granulomatous disease. In principle, the major group of monogenic disorders that could be treated by *ex vivo* therapy are the haemoglobinopathies. Gene therapy for this group of diseases is complicated by the mechanisms that control the expression of both the α-globin and β-globin gene cluster. Expression of genes in these clusters requires the action of LCRs that operate at a considerable distance from the genes controlled (see Chapter 4). Increased understanding of these processes has led to recent advances in this area (see Further reading).

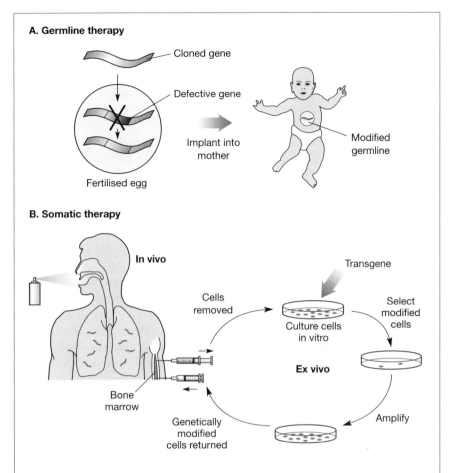

A. Germline therapy

Cloned gene

Defective gene

Fertilised egg

Implant into mother

Modified germline

B. Somatic therapy

In vivo

Cells removed

Culture cells in vitro

Transgene

Select modified cells

Ex vivo

Bone marrow

Genetically modified cells returned

Amplify

Figure 6.1 Germline and somatic gene therapy. Germline gene therapy (top) requires that the transgene is introduced into the fertilised egg or into the embryo before the germline separates from the soma. As a result, the baby is born with modified somatic and germline cells and will pass the modification on to the next generation. In somatic gene therapy the change is limited to the cells of the subject's body and is not passed on to the next generation. Somatic gene therapy can be *ex vivo* or *in vivo*. *Ex vivo* therapy involves removing cells from the patient, in this case blood cell progenitors from the bone marrow, and modifying them *in vitro*. After the modification, the cells may be cultured to increase their number before reintroduction into the patient. *In vivo* therapy involves direct introduction of the transgene into the cells of the subject's body. In this case a transgene is introduced into the cells that line the lung by means of an aerosol.

Clearly, *ex vivo* approaches for diseases such as CF and muscular dystrophy will not be practical. The cells of the lung epithelia divide only very slowly so it is not possible to culture them *in vitro*. Even if it were, it is not easy to imagine how the lungs of affected individuals could be readily repopulated by cells manipulated *in vitro*. The accessibility of such cells thus becomes a major factor in determining whether gene therapy is possible. In the case of diseases such as CF it will be necessary to manipulate the cells *in vivo*. This requires that the transgene is delivered to those cells affected by the disease. Thus DMD would be treated by introducing the dystrophin gene into muscle cells, CF by introducing the *CFTR* gene into the cells of the airway, and so on. In both diseases these are the cells whose malfunction causes the major life-threatening effects. Nevertheless, in both examples there are many other cell types that fail to function normally and contribute to the pathophysiology of the disease. CF patients may suffer, *inter alia*, from pancreatic insufficiency, gastrointestinal malformation, defects in the vas deferens and liver problems. DMD patients may also have heart and central nervous system defects. Even if the defects were successfully corrected in the tissues primarily affected by the disease, the disease may not be fully cured. Furthermore, many somatic cell types may not be easily accessible for treatment.

Gene addition or replacement

The most perfect form of gene therapy would be to completely replace the defective gene with a functioning transgene. This would allow therapy for all types of mutation including dominant ones. Replacement would involve homologous recombination between the transgene and the chromosomal copy. This is possible as there are many situations where geneticists rely on such recombination, for example in the generation of gene deletions in transgenic mice. However, gene replacement requires procedures to select a small minority of cells that have the alteration required. The selection procedures make use of sensitivity or resistance to drugs. *In vivo* gene therapy would need to be carried out without such selection. Although gene replacement has been shown to occur at a very low frequency in experiments with mouse embryonic stem cells, such procedures in human cells will not be technically feasible in the foreseeable future. Germline therapy will require accurate gene replacement, because this is the only way to be sure that the gene will be expressed in a completely normal fashion. This is one of the powerful practical reasons that precludes germline therapy for the present.

For somatic gene therapy, this leaves gene addition rather than replacement, where the transgene works alongside the mutated gene in the same cell. This should allow correction of the lack of gene function that occurs in recessive conditions. However, it cannot correct a dominant mutation because the mutant protein will still be produced and exert its detrimental effect. It is theoretically possible to use antisense strategies to inhibit a gene affected by a dominant mutation. This relies on producing an RNA molecule, or introducing an oligonucleotide, that is complementary to the mRNA originating from the gene. The mRNA and this **antisense mRNA** form a double-stranded hybrid that cannot be translated and is specifically degraded. Targeting to a mutant mRNA can be achieved by using an oligonucleotide whose sequence is complementary to mRNA from the mutant version of the gene.

The transgene can be integrated into a chromosome or can exist as an independent copy (**episome**). The advantage of the former is that the transgene will be stable in cell division and be inherited by progeny cells. The disadvantage is that, as noted above, the site of integration cannot be controlled. It is possible that the new gene will insert within a functioning gene, resulting in its inactivation. Alternatively it may activate an oncogene by integrating next to it. As a consequence the cell may become transformed into a cancer cell. This is not a theoretical speculation. Some retroviruses are known to be oncogenic by this mechanism.

Animal models

Development of gene therapy techniques will require a large amount of experimentation. For obvious reasons it is difficult to do most of this on human subjects. Animal models are extremely useful to demonstrate the principle and develop the technology. Some diseases have natural counterparts in animals. For example, *mdx* in mice and *xmd* in dogs (golden retrievers) are X-linked mutations that cause a disease similar to DMD. Alternatively, transgenic mice can be generated with the homologue of the human disease gene deleted. Usually these mice suffer from similar symptoms and can therefore be used to develop gene therapies. It is important to realise that there will also be important differences and the results may not always be directly transferable from the animal model.

Table 6.1 Advantages and disadvantages of the various approaches used to transfer genes.

	Advantages	Disadvantages
Retroviruses	Integration results in stable modification of target cell Infect replicating cells Suitable for *ex vivo* treatment	Uncontrolled integration may have oncogenic consequences Can not infect non-dividing cells Provoke immune response
Adenoviruses	Infect non-dividing cells, so suitable for CF and DMD Non-integration avoids safety hazard of uncontrolled integration Efficient gene transfer, at least *in vitro*	Expression of transferred gene is transient Provoke strong immune reaction
Lipoplexes	Non-immunogenic, so safe Can carry large DNA molecules	Inefficient delivery Transient expression
Naked DNA	Non-immunogenic Simple to prepare and administer to muscle cells	Inefficient gene transfer Limited in cell types that can be treated Expression transient

6.3 Methods for transferring transgenes into target cells

Gene therapy relies upon methods of introducing transgenes into target cells. This means a **vehicle** is needed for its delivery. Sometimes the term 'vector' is used in this context. In this account, the word 'vector' is reserved for its normal meaning of a DNA molecule to which a gene may be spliced to mediate its expression and/or replication in the host cell. Sometimes the use of the two words will overlap, for example a retroviral genome into which a transgene has been spliced is a vector, but the **virion** (see below) containing the recombinant genome is a vehicle. The utility of the vehicle will depend on a number of factors:

- efficiency of delivery to the target cell;
- specificity of delivery for the target cell (**tropism**);
- whether the target cell needs to be dividing;
- whether the vehicle will provoke an immune response;
- the size of the DNA that can be carried;
- stability and longevity of the gene in the target cell;
- expression of the introduced gene.

Various approaches can be taken to transfer the gene to the target cell. The advantages and disadvantages of these are summarised in Table 6.1 and are discussed in more detail in the following sections.

6.3.1 Virus-based vehicles for gene therapy

Retrovirus-based gene transfer

Retroviruses (Figure 6.2) have two identical RNA genomes. Upon infection the RNA is copied into DNA by a virally encoded reverse transcriptase. This DNA

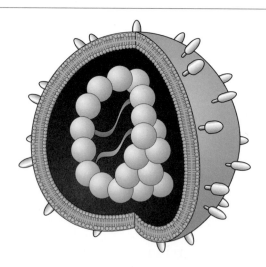

Figure 6.2 Retrovirus virion. The mature virus particle found outside the cell is called a virion. The retrovirus virion is composed of an outer lipid envelope into which are inserted envelope proteins that recognise specific proteins on the outside of target cells. Within the outer lipid layer is an inner nucleocapsid, a hollow protein structure made up of the gag and pol proteins surrounding two RNA genomes.

copy is then integrated into the host genome in a process that depends upon the **long terminal repeats** (LTRs) at each end of the genome (Figure 6.3). The retroviral genes are transcribed from a promoter within the 5' LTR regions. The resulting RNA molecules are packaged into the virus particles. This packaging process requires a sequence called ψ. As well as the *pol* gene encoding reverse transcriptase, the retroviral genome contains the *gag* gene that encodes a viral core protein and the *env* gene that encodes a viral envelope protein. The *env* protein recognises receptors on target cells, facilitating infection.

Figure 6.3 shows how retroviruses may be used as gene delivery systems. Because retroviruses can be oncogenic if they integrate upstream of a cellular oncogene, it is essential that any retrovirus-based gene therapy product is not contaminated by replication-competent viruses. This is achieved by using a packaging cell line in which essential retroviral genes are divided between two different DNA molecules, neither of which contains the ψ packaging sequence necessary for incorporation into the virion (the mature virus particle). A third DNA molecule carries the gene to be transferred together with a functioning ψ sequence sandwiched between LTRs. This packaging cell line produces retrovirus particles containing the gene of interest that can infect the target cell. Once inside the target cell, the pol protein contained within the virus particle will ensure that the gene is integrated into the host genome. The LTRs are necessary for integration and also act to express the gene once integrated. Because multiple recombination events will be required to produce a functioning retrovirus in the packaging cell line, it is highly unlikely that the preparation will be contaminated. Nevertheless, each batch produced for clinical testing is carefully examined for any replication-competent virus particles.

The advantages of retrovirus-mediated gene transfer are its efficiency compared with other methods of gene transfer, together with the stability of the introduced gene. It is generally used for *ex vivo* gene therapy, which is based on replicating cells. Many gene therapy trials are currently in progress using retroviral vehicles for gene transfer. Nevertheless, there are significant problems with the use of retroviruses for gene therapy.

1. Retroviruses can only infect proliferating cells because the viral genome integrates into the host chromosome and can only gain access to the chromosomes when the nuclear envelope disappears at mitosis. This prevents the use of retroviral vehicles for the treatment of genetic diseases such as DMD or CF where the cells most affected are non-dividing or only dividing slowly.
2. Retrovirus-mediated gene transfer is limited in the size of DNA molecule that can be accommodated, which mostly limits its use to cDNA.
3. Retrovirus-mediated gene transfer does not specifically infect one cell type. This results in a loss of efficiency as most of the genes do not reach the target cells. Moreover, when it is used to alter the characteristics of a particular cell type, such as a cancer cell, modifying the wrong cell type may be actively counter-productive. One approach for improving target specificity is to modify the viral env protein that interacts with receptors on the exterior of the host cell. This can be done by fusing a protein hormone to the env protein so that the virus may be targeted to cells

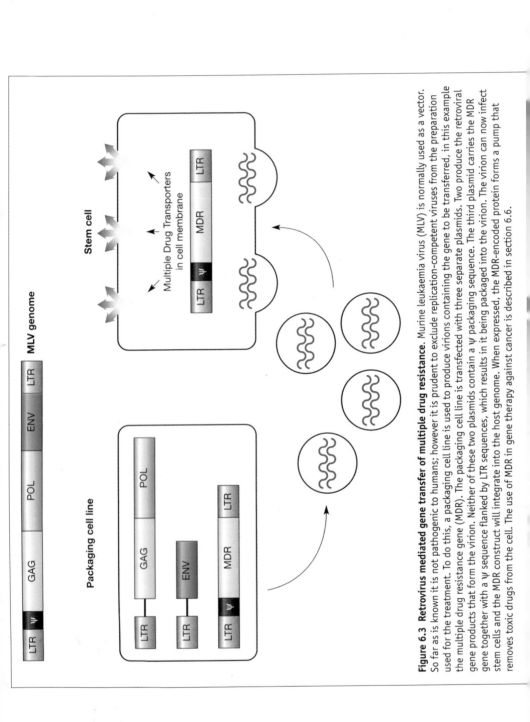

Figure 6.3 Retrovirus mediated gene transfer of multiple drug resistance. Murine leukaemia virus (MLV) is normally used as a vector. So far as is known it is not pathogenic to humans; however it is prudent to exclude replication-competent viruses from the preparation used for the treatment. To do this, a packaging cell line is used to produce virions containing the gene to be transferred, in this example the multiple drug resistance gene (MDR). The packaging cell line is transfected with three separate plasmids. Two produce the retroviral gene products that form the virion. Neither of these two plasmids contain a ψ packaging sequence. The third plasmid carries the MDR gene together with a ψ sequence flanked by LTR sequences, which results in it being packaged into the virion. The virion can now infect stem cells and the MDR construct will integrate into the host genome. When expressed, the MDR-encoded protein forms a pump that removes toxic drugs from the cell. The use of MDR in gene therapy against cancer is described in section 6.6.

displaying the receptor for that hormone. Another strategy is to replace the env protein with that of another virus. For example, in one experiment the mouse leukaemia virus (MLV) env protein was replaced with the env protein of human vesicular stomatitis virus (HSV). The engineered MLV virus only infected cells displaying the HSV receptor and no longer infected the normal target cells of MLV.

4. Retroviruses integrate at random sites in the genome. If they integrate near a cellular oncogene they may cause its activation through transcriptional activation.

The first successful use of gene therapy in treating severe combined immunodeficiency (SCID) was based on an *ex vivo* regime using a retroviral vector. One cause of SCID is an inherited lack of ADA, which particularly affects lymphocytes. Two ADA-deficient children were treated by *ex vivo* manipulation of isolated lymphocytes. ADA was introduced into these cells by two different routes: firstly, by introducing the ADA gene via a retrovirus-based vehicle and, secondly, by introducing the enzyme itself encapsulated by polyethylene glycol. The treatment, which was repeated every 6 weeks, resulted in marked clinical improvement, but it is unclear whether this was due to the introduced DNA or protein, especially as no ADA gene expression could be detected in one of the children.

Adenovirus-based gene transfer

Adenoviruses (Figure 6.4) are double-stranded DNA viruses that naturally infect the non-dividing cells of the respiratory and gastrointestinal tracts.

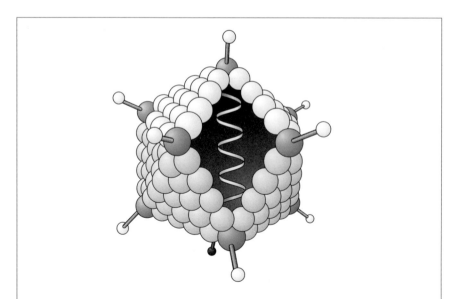

Figure 6.4 An adenovirus virion. This is an icosahedral structure consisting of a **capsid** made up of 12 different proteins enclosing a single double-stranded DNA molecule 35 kb in size.

They make an attractive vehicle for gene therapy because they have evolved to evade host immune mechanisms and to be highly efficient at infecting their target cells. Furthermore, once inside the target cell they replicate as episomes and so avoid the dangers of uncontrolled integration. There are at least 50 different **serotypes**. Serotypes 3 and 5 show a high degree of tropism for the respiratory tract and thus these strains have been used as vehicles for CF gene therapy. The ability of adenoviruses to infect non-proliferating cells has resulted in attempts to use them for gene therapy of DMD, where it is necessary to treat non-dividing muscle cells.

Of course it is necessary to disable the adenovirus to prevent it replicating and killing the host cell. Gene expression can be divided into early and late phases. DNA replication and expression of the genes whose products constitute the virion depend upon prior expression of early genes. The infective cycle can therefore be halted by deleting E1, an essential early gene (Figure 6.5). At the same time this makes room for the insertion of foreign DNA, up to a size of about 6 kb.

Adenoviruses are efficient at delivering DNA to their target cells. However, a major drawback is that clinical trials have shown that even disabled forms provoke a strong immune response. This limits the dose that can be safely administered. Moreover, expression is transient so that the therapy would need to be repeated. Repeated therapy results in the immune response becoming more severe. Recently, adenovirus vehicles with deletions of other early genes such as E2, E3 and E4 (Figure 6.5) have been constructed to further reduce late gene expression which might be responsible for the immune response.

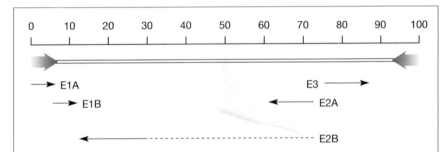

Figure 6.5 Early gene transcription in adenoviruses. The adenovirus genome is 36 kb in size conventionally represented by a map with 100 units. Each end of the molecule consists of a 100 bp sequence that is an inverted repeat of the sequence at the other end. During the first 6–8 hours after infection, mRNA molecules representing about 25% of the genome are derived from scattered regions of the genome called E1 to E4. Each region gives rise to multiple transcripts by differential splicing but all have the same 5′ end. The resulting overlapping mRNA molecules give rise to proteins that are subsets of one another. In some cases one mRNA molecule is read in overlapping reading frames so that the resulting proteins have common amino-terminal sequences but different carboxy-terminal sequences. The very first region to be transcribed is E1A; the E1A protein is required for the expression of all other adenovirus genes. Early gene transcription allows DNA replication followed by late RNA transcription that encode the capsid proteins of the virion.

Adeno-associated viruses

Adeno-associated viruses (AAV) are naturally replication-defective viruses that depend on other helper viruses, such as adenovirus, to provide essential functions. AAV-based systems offer a number of advantages over other virus systems: they are not naturally pathogenic, they are tropic for airway cells, they infect non-dividing cells, they show longer-lasting expression (up to 6 months in animal models) and they can show chromosome-specific integration in chromosome 19. Although not all of these theoretical advantages have been borne out in practice, several trials for CF gene therapy have been undertaken with this vector.

6.3.2 Non-viral methods of gene transfer

Liposome- and lipoplex-mediated gene transfer

In aqueous solution DNA molecules are negatively charged. The outside of most cells are also negatively charged, so DNA molecules and cell surfaces may be expected to be mutually repellent, which is thought to limit the efficiency of DNA uptake by cells. This problem may be overcome by encapsulation of the DNA in **liposomes**. Liposomes are spheres consisting of lipid molecules surrounding an aqueous interior. The lipid molecules used contain hydrophobic and hydrophilic domains. They form a bilayer in which the hydrophilic domains face outwards towards the surrounding aqueous environment and inwards towards the water-filled interior. DNA molecules can be encapsulated within liposomes. However, liposomes are typically 0.025–0.1 μm in size but plasmids are typically over 2 μm in size. Thus only a small number of plasmids can be fitted into each liposome and as a result liposomes are very inefficient vehicles for gene transfer.

This problem was overcome with specially formulated lipids where the hydrophilic domain is positively charged (cationic), which attracts both the negatively charged DNA and the negatively charged cell surface. These cationic lipids were very efficient at encapsulating DNA. The resulting lipid–DNA complexes are much more complicated than liposomes. The plasmids are encapsulated in tube-like structures, which in some ways resemble the lipid envelope that surrounds the protein capsid of some viruses. Because these structures are so different from liposomes they were given a new name, **lipoplexes**. The main advantage of lipoplex-mediated gene transfer is that lipoplexes are non-immunogenic so they are safer than virus-based methods. In addition, lipoplexes are easy to prepare and are not limited in the size of DNA molecule that can be carried. The main disadvantage is that the efficiency of transfer is lower than virus-based methods. Nevertheless recent trials have shown that in some cases they can be as efficient as virus-based vehicles.

Naked DNA

During the course of experiments designed to test the efficiency of various liposome/lipoplex formulations, naked DNA was used as a negative control since, theoretically, naked DNA should not be taken up by cells.

Surprisingly it was found that naked DNA injected directly into the muscle tissue of mice was taken up and expressed by muscle cells. This is thought to occur by DNA entering the cell through small lesions in the cell membrane. This opens up a novel route for the introduction of genes into muscle cells. One possibility is to use muscle cells to produce proteins whose action is not cell-limited, such as insulin or blood clotting factors. As discussed below, another idea is the use of naked DNA to express proteins that would act as vaccines against infectious diseases or even cancers.

6.4 Gene therapy for cystic fibrosis

Soon after the *CTFR* gene was cloned it was demonstrated that the wild-type gene could correct the chloride conductance defect of a CF cell line *in vitro*, demonstrating that in principle it should be possible to treat the disease by gene therapy (see Chapter 4). The life-threatening consequences of the disease affect the epithelial cells that line the lung airways and thus these are the cells into which the wild-type gene would have to be introduced. Clearly, it will not be possible to treat the cells by *ex vivo* therapy, so an *in vivo* method of treatment will need to be developed.

The first experiments were carried out with an adenovirus vector using transgenic mice lacking *CFTR* function as a model (CF transgenic mice). They showed that the gene could be delivered to the target cells in the airway and correct the chloride ion transport defect, for 6 months after introduction of the wild-type gene in some experiments. The scene was set for gene therapy trials with human CF patients. There are several methodological problems that need to be considered before describing the results.

Firstly, in which part of the airway is expression most important? *CFTR* expression is highest in the submucosal gland, which may secrete water on to the airway surface. However, *CFTR* is also expressed in the epithelia that line the alveoli, so CFTR function may be important there as well. It may be that restoration of function in airway progenitor cells may be necessary for successful treatment. At present, gene delivery systems are based on aerosols, which can only access epithelial cells and not submucosal cells or progenitor cells

Secondly, what proportion of cells need to be corrected for relief of the clinical symptoms? *In vitro* experiments with monolayers of CF epithelial cells suggest that restoration of chloride ion conductance in only 6% of the cells is sufficient because chloride ions can move laterally from cell to cell through **gap junctions**, which connect adjacent cells. However, sodium ion absorption is cell-limited; thus most cells will need a functioning CFTR protein if regulation of sodium ion absorption is important.

Thirdly, how should preliminary trials be conducted? If the gene is delivered to cells deep in the lungs it will be difficult to gain access to these cells to measure whether the gene has been delivered and whether there is any evidence of *CFTR* function. Alternatively, the gene could be delivered to the nose where it is possible to sample cells to determine whether the

CFTR gene has been delivered and, if it has, to monitor its expression and function. However no clinical outcome can be observed because the nasal passages are not involved in the pathophysiology of CF. A compromise may be the maxillary sinus, which can be accessed to monitor the success of gene delivery and function but which is also clinically relevant.

Adenovirus trials

There have been a number of clinical trials based on adenovirus vehicles, but with only limited success in gene delivery and function. *CFTR* expression was variable and transitory. Increasing the dose of the virus to improve the outcome results in a strong immune response in the form of inflammation and fever. The problem may be that the cells that need to be treated are the columnar ciliary cells that line the airway. Previous *in vitro* experiments have been carried out with the underlying basal cells. The columnar cells are much more resistant to infection by adenoviruses. Perhaps this should not be surprising. Adenoviruses were chosen as vehicles because they have evolved to infect the lung. While this may be true, it is probably equally true that columnar cells have also evolved to be resistant to such infection.

Lipoplexes

Trials where aerosols of lipoplex-encapsulated DNA were sprayed into the nasal cavity have had some very limited success. Gene transfer was monitored by the presence of *CFTR* DNA and mRNA, and *CFTR* expression was monitored by chloride ion conductance and sodium ion absorbance. Both transfer and expression were detected, but the levels were low and did not persist for any significant length of time. The levels were comparable to those observed with adenovirus vehicles but, importantly, without the inflammatory response. These trials were conducted using the nasal epithelium, which is not medically relevant to CF and is probably an easier target than the usually infected lower airways typical of CF patients. Moreover, although the degree to which it is necessary to remediate *CFTR* function is not known, the levels of *CFTR* expression observed are not likely to prove sufficient for clinical improvement.

Lipoplexes are taken up into the cell by **endocytosis**: the cell membrane invaginates to form an intracellular vesicle. These vesicles fuse to form structures called **endosomes**. A problem with lipoplexes is that they remain trapped in the endosome and so the DNA does not reach the nucleus. A suggestion to overcome this problem is to encapsulate an adenovirus vector in a lipoplex. This may combine the benefits of both vehicles: the negatively charged lipoplex provides an efficient means of access into the cell, while the adenovirus can escape from the endosome and deliver the DNA to the nucleus.

Future directions

Although it is exciting that clinical trials for CF have been undertaken with some positive results, clinically effective treatment is still a long way off. Far more efficient ways have to be found to deliver and express the *CFTR* gene. One exciting development is the construction of human artificial chromosomes based on an α-satellite sequence as a centromere. This may provide a vector for the safe and permanent maintenance of the *CFTR* gene. Since such a vector will be capable of carrying a large DNA insert, it might be possible to transfer the entire *CFTR* chromosomal locus instead of *CFTR* cDNA. This may enable the pattern of expression to match the natural system more faithfully.

A second way forward may be treatment *in utero*, where immune tolerance to adenovirus may be greater. It may also avoid the early and possibly irreversible damage to the lung that occurs in CF patients. One intriguing report shows that in a CF transgenic mouse, transient expression of *CFTR in utero* results in a permanent reversal of the CF phenotype. The *CFTR*-deleted mice used in the study suffer from abnormalities of the gastrointestinal tract that mimic meconium ileus, a form of intestinal blockage seen in 5–10% of newborn CF patients. Transient expression completely prevented this abnormality so *CFTR* expression may only be required briefly during development, but not subsequently. If this is true, it may be necessary to reassess current strategies for gene therapy of CF. However, some caution would be appropriate before such a conclusion is drawn. At present it is not known how relevant this model will be to the human airway. Previous experience has shown that studies with the CF mouse model do not always transfer to the human situation.

6.5 Gene therapy for Duchenne muscular dystrophy

Gene therapy for DMD has been successfully achieved in transgenic mice, but has not reached the stage where human trials can be attempted. Several problems specific to DMD need to be solved. Firstly, ways have to be found to access muscle cells to effect gene transfer. Secondly, the extreme size of the dystrophin gene (14 kb) makes it difficult to incorporate the wild-type gene into adenovirus vehicles, which have an upper limit of 6 kb.

The *mdx* mouse shows similar defects to the human disease. Human dystrophin cDNA has been transferred to muscle cells using an adenovirus vector either directly injected into muscle tissue or injected into the heart and allowed to spread around the body with the bloodstream. The size problem has been partially solved by making use of a mini-dystrophin gene, which originated in a patient with BMD. The deletion responsible resulted in a reduction in the size of the cDNA to 6.3 kb. The reading frame was maintained and the resulting dystrophin protein retained considerable function, even though it was only half the size of the wild-type protein. Expression of this mini-gene in *mdx* mice results in expression of dystrophin, which is correctly located within the muscle cells and completely corrects the pathological phenotype. Retroviral vehicles have also been used. Because they only infect replicating cells, the gene was transferred to embryonic **myoblasts**. Myoblasts can be reimplanted *in vivo*, but unfortunately expression of the dystrophin gene is transient when this is done.

One approach to gene therapy for DMD is based on the observation that even in patients with a severe form of the disease a small amount of normal or nearly normal dystrophin protein is often detected. This is thought to arise as follows. Most severe DMD mutations destroy the reading frame of the dystrophin-encoding mRNA. During RNA processing, mistakes can occur that result in an exon carrying the mutation being omitted. In a small number of mRNA molecules this can restore the reading frame and result in the production of dystrophin. The splicing process can be redirected using synthetic oligonucleotides. Experiments with cells *in vitro* have shown that it is possible to restore the reading frame of a mutated dystrophin mRNA using such an approach.

6.6 Gene therapy for non-heritable disorders

Gene therapy can also be used for non-heritable disorders, and these applications are a lot nearer introduction into medical practice than gene therapy for heritable diseases. In reality, they form the bulk of gene therapy trials in progress.

Cancer therapies
Many gene therapy protocols are being investigated as possible treatments for cancers. One idea is to deliver a gene to a target cell whose product will activate a prodrug, which then kills the manipulated cell but does not affect the rest of the cells in the body. For example, the thymidine kinase gene from herpes simplex virus can be delivered to brain tumour cells using a retroviral vehicle (Figure 6.6).

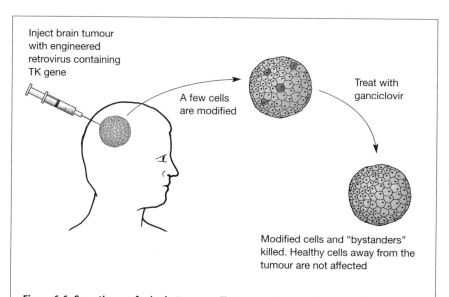

Figure 6.6 Gene therapy for brain tumours. The tumour is directly injected with a retrovirus containing the mouse thymidine kinase (TK) gene and a few cells take up the vector, shown in blue. These cells convert the prodrug ganciclovir into an active form and are killed (grey cells). Because of the bystander effect surrounding cells are also killed.

The engineered retrovirus is injected directly into the brain tumour and the patient is then treated systemically with ganciclovir. This drug is an analogue of guanosine. It is phosphorylated by thymidine kinase to the triphosphate form, which can be incoporated into DNA. Incorporation inhibits further DNA replication and kills the cells. Thus ganciclovir is cytotoxic but only when phosphorylated by thymidine kinase. Although only a small proportion of the cells express thymidine kinase, a larger number of surrounding cells are killed. This is known as the bystander effect. The reason why this happens is not completely understood, but it is thought that the toxin generated in one cell can spread to adjacent cells through gap junctions.

Retroviruses can only replicate in dividing cells but brain cells are non-dividing. The use of retroviruses thus targets the treatment to tumour cells which will be the only dividing cells in the brain. When this treatment was used in phase I clinical trials a further unexpected benefit was observed: secondary tumours elsewhere in the body showed signs of regression. It is thought that lysis of cells in the primary tumour released a high concentration of tumour-specific antigens, which provoked an increased immune response against the secondary tumours.

Another possible use for gene therapy against cancer is to protect blood haemopoietic stem cells from the effects of chemotherapy. The usefulness of chemotherapy is often limited by two factors. Firstly, resistant lines arise that express elevated levels of the multiple drug resistance (MDR) protein, which pumps toxic drugs out of the cell. Secondly, haemopoietic stem cells naturally express very low levels of MDR and are killed by low concentrations of toxic drugs. This leads to a condition known as leucopenia, a lack of white blood cells, and the patient becomes immunocompromised. The problem of resistance could be overcome by increasing the initial concentration of drugs to kill the cancer cells before resistant lines arose. However, such a course of action would exacerbate the second problem and kill the patient. A solution to this impasse is to use gene therapy to increase the expression of MDR in haemopoietic cells. The way a recombinant retrovirus can be used to elevate the level of MDR expression is described in Figure 6.3.

Haemopoietic stem cells can be efficiently transduced by retroviruses using *ex vivo* therapy and can then repopulate bone marrow after reimplantation. Haemopoietic stem cells express a surface antigen called CD34 and thus may be selected using an anti-CD34 antibody. There are sufficient CD34$^+$ cells in peripheral blood for the treatment so there is no need for painful bone marrow sampling. They are induced to grow *in vitro* using a cocktail of growth factors that stimulate their proliferation and are then transduced with the retrovirus carrying the MDR gene. Meanwhile the patient is subjected to aggressive chemotherapy for 3–4 days with drugs such as taxol. The manipulated stem cells are then reimplanted along with unmanipulated marrow cells. In phase I clinical trials of this treatment one patient showed MDR transfer in CD34$^+$ cells 3 months after treatment.

There is much research directed towards correcting the somatically acquired mutations that give rise to cancer. Dominantly acting oncogenes such as activated *ras* may be inhibited by antisense mRNA that prevents translation of the oncogene mRNA. Over half of cancers have mutations in tumour suppressors such as RB and p53 (see Chapter 4). When wild-type genes are introduced into cell lines from tumours carrying these mutations the tumour characteristics disappear. If these genes could be introduced into tumour cells *in vivo* then cells could either reacquire normal control of cell regulation or, if the genomes are irreversibly damaged, p53 may induce apoptosis (programmed cell death). The p53 protein functions as a dimer. Some p53 mutant proteins interfere with the assembly of normal subunits into dimers. Such mutations are said to be **dominant negative mutations**. Clearly, introducing a wild-type copy will not restore p53 function and it is necessary to turn off the mutant gene. This may be achieved by antisense oligonucleotides that bind to the normal mRNA and prevent translation.

One important defence against cancer is the immune system. Several trials have been designed to recruit the immune system to help fight the cancer. One idea of how this could be achieved is to programme cancer cells to express non-self HLA markers on the cell surface. This should result in the cells being destroyed by cellular immunity mechanisms. In one experiment, lipoplexes containing the gene encoding the major histocompatibility antigen HLA-B7 were injected into malignant melanoma tumours. In one-third of the patients the tumour shrank or even disappeared. Encouragingly, there were signs that even tumours that were not injected started to regress. This is important because malignant melanoma spreads through the body and it would be difficult to treat all tumours separately.

A second way to enlist the immune system to fight a cancer is to increase its sensitivity to cancer cells. The immune system is thought to be an important natural line of defence against cancer. It is capable of recognising cancer cells and killing them, but in an established cancer is clearly failing to do this. The idea of stimulating the immune system to fight cancer is an old one. In the early part of this century an American surgeon called William Coley noticed that the tumours of some patients regressed after a severe bacterial infection. From this he developed a cancer treatment based on deliberate bacterial infection. As might be expected, this did not turn out to be very successful because the infection could not be controlled. Coley modified the treatment by injecting a preparation of dead bacterial cells called Coley's toxins. This treatment was effective in some cases but was never consistently successful and after a while it fell from use. The reason for the original observation was later shown to be that the bacterial infection stimulated the production of a **cytokine** called tumour necrosis factor (TNF) that stimulated cell-mediated immunity. The production of recombinant TNF led to hopes that a modern version of the old bacterial infection treatment could be used against cancer, this time using TNF to stimulate the immune system. Unfortunately, TNF has severe side-effects, such as the induction of fevers and general malaise; once again the treatment failed and for very similar reasons.

The latest version of this treatment is to target the production of cytokines to the cancer cells themselves. This is done by targeting genes encoding cytokines to the cancer cells. This stimulates the immune system, but only in the vicinity of the tumour. The attraction of this approach is that it is not necessary to target every cancer cell or even every tumour where there are multiple secondary tumours. By stimulating the immune system to destroy a few cancer cells, its interest in all cells carrying the same tumour antigen is aroused.

DNA vaccines

A different area that is attracting a lot of interest is the use of naked DNA as a novel type of vaccine. The idea here is to induce muscle cells to produce antigens from infectious agents or tumour-specific antigens. A particular advantage of this strategy is that as the antigen is expressed by cells it may trigger cellular as well as humoral immunity. The principle has been shown to be effective for a variety of infectious diseases in mice. Clinical trials are underway in humans for influenza vaccination using this strategy. It is possible that trials for HIV, herpes and malaria may start in the near future. In the longer term, it is hoped to use proteins commonly found on the surface of cells in some cancers to produce vaccines against these cancers.

6.7 Summary

- Gene therapy seeks to treat genetic disease using the unmutated version of the gene affected.

- Two strategies can be employed: somatic gene therapy, which limits any genetic change to the lifetime of the person treated, and germline therapy, which affects gametes and any changes will thus be inherited by future generations. Germline therapy is currently prohibited for ethical reasons.

- Gene therapy can act by replacing the mutated gene or adding a wild-type version to work alongside it. Gene replacement requires techniques that are not yet available, so gene therapy of inherited diseases is limited to recessive mutations.

- Cells may be modified by *ex vivo* techniques, where cells are removed from the body, genetically modified and then reintroduced. *Ex vivo* therapy is suitable for modification of blood stem cells.

- Diseases such as DMD and CF must be treated by *in vivo* therapy, where the gene is introduced to the cells within the body of the patient.

- Various vehicles are used to introduce genes into cells:
  - Retroviruses introduce genes into replicating cells and are used for *ex vivo* therapy.
  - Adenoviruses are tropic for airway cells and are efficient at carrying genes into cells, but provoke a strong immune reaction in the patient.
  - Lipoplexes encapsulate the DNA into cationic lipids. They do not induce an immune reaction, but may be less efficient than other methods.
  - DNA can be directly injected into muscle cells. This is being used for DNA-based vaccines.

- The first trials for CF gene therapy have successfully demonstrated that genes can be delivered to cells *in vivo*. However, expression was transient and not sufficiently strong to cause a clinical improvement. Adenovirus-based therapy provoked a strong immune reaction.

- Gene therapy for DMD has been demonstrated in a mouse model but so far there have been no clinical trials.

- Most gene therapy research is currently focused on the treatment of non-inherited diseases. Various ways of using gene therapy to combat cancer have been tried and some have shown encouraging results in clinical trials.

Further reading

General
VARIOUS AUTHORS (1997) Special report: making gene therapy work. *Scientific American*, **276** (6), 79–103.

Cystic fibrosis
BOUCHER, R.C. (1996) Current status of CF gene therapy. *Trends in Genetics*, **12**, 81–84.

CAPLAN, N.J. AND ALTON, E.W. (1996) Gene therapy for cystic fibrosis. *Chemistry and Industry*, **8**, 290–292.

WAGNER, J.A. AND GARDNER, P. (1997) Towards cystic fibrosis gene therapy. *Annual Review of Medicine*, **48**, 203–216.

Duchenne muscular dystrophy
DICKSON, G. (1996) Gene therapy of Duchenne muscular dystrophy. *Chemistry and Industry*, **8**, 294–297.

FASSATI, A., MURPHY, S. AND DICKSON, G. (1997) Gene therapy of Duchenne muscular dystrophy. *Advances in Genetics*, **35**, 117–149.

Haemoglobinopathies
EINERHAND, M.P.W., ANTONIOU, M., ZOLOTUKHIN, S. *et al.* (1995) Regulated high level human β-globin expression in erythroid cells following adenovirus mediated gene transfer. *Gene Therapy*, **2**, 336–343.

Gene therapy for cancer
VILE, R.G. (1996) Gene therapy for treating cancer: hope or hype? *Chemistry and Industry*, **8**, 285–289.

Gene testing

Key topics

- When gene testing is used
- Types of mutation
- General principles of testing
- Testing for mutations in the *CFTR* gene
- Testing for mutations in the dystrophin gene
- Testing for uncharacterised mutations
  - ○ Single-stranded conformational polymorphisms
  - ○ Denaturing gradient gel electrophoresis
  - ○ Mismatch cleavage detection
  - ○ Protein truncation test
  - ○ DNA chips
  - ○ Chromosome tracking

7.1 Introduction

The advances in gene isolation and characterisation described in this book enhance enormously the ability to diagnose genetic disease by the more traditional methods based on a physician's skill supported by biochemical tests. This chapter is concerned with the techniques that can be used routinely to detect mutations in genes of interest. The application of these techniques often raises significant ethical problems; these are considered in Chapter 11.

There a number of different situations where gene testing might be carried out.

- Where there is a known risk of a particular disease, embryos may be tested with a view to termination; alternatively, prospective parents may wish to know their carrier status in order to make reproductive decisions. The optimum situation in embryo testing is where the nature of the mutation has previously been defined, thus allowing its presence to be specifically tested in the embryo. In an autosomal recessive disorder this will require two mutations to be identified. In the absence of such information the chromosome(s) harbouring the mutation may be tracked by linked polymorphic markers.

● Neonatal population screening for genetic diseases where there is a clear benefit derived from early diagnosis and there is a simple and reliable test available. Neonatal screening for phenylketonuria is common using the Guthrie 6-day blood-spot test. A small sample of blood is taken from the baby and preserved by drying on to a card. Phenylalanine levels in the blood spot are then assayed. However, the same blood spot can provide material for other tests, such as screening for the ΔF508 *CFTR* mutation in European populations.

● To confirm or rule out a diagnosis of genetic disease made on the basis of medical symptoms.

● To diagnose the status of an individual, normally an adult, at risk from a late-onset genetic disease such as HD or one of the inherited cancer predispositions.

● To determine the status of genes predisposing to complex diseases. This area is likely to grow in importance as tests are developed for susceptibility to diseases such as IDDM.

● Where a candidate gene involved in a genetic disease has been identified, genetic testing is used to search for a correlation between the presence of a mutation and the onset of disease symptoms. This is often the final step in the hunt for a disease gene.

DNA testing has been revolutionised in the last few years
A few years ago work in a DNA diagnostic laboratory would mainly consist of tracking disease chromosomes through a family using RFLPs analysed by Southern blotting. Southern blotting is laborious, involves the use of radioisotopes, with the attendant safety problems, and requires significant amounts of biological material. RFLP analysis required extensive work to establish the phase of mutations and linked RFLP polymorphisms segregating in a family. Many meioses were uninformative because the RFLPs were not heterozygous, and if the linked markers were not sufficiently close to the mutation the diagnosis could be falsified by recombination.

Techniques used to screen for mutations have been revolutionised in the last few years by two developments. Firstly, cloning the disease genes has allowed tests to be directed towards the gene sequences themselves rather than tracking the inheritance of mutant chromosomes through families. Secondly, PCR is used to amplify DNA sequences so that enough product is generated to be visualised by ethidium bromide fluorescence or silver staining. Chromosome tracking is still sometimes required when the nature of the mutation is unknown, but this is now carried out using highly informative microsatellite polymorphisms that can be analysed by PCR.

Three different challenges to the gene tester

In some cases the nature of the mutation likely to affect a gene is known. Firstly, where only one or a small number of different mutations in a population are responsible for a disease; typically, this results from founder mutations. Secondly, where the disease commonly occurs because of a particular type of mutation. The obvious example of this would be TRE mutations, but we may also consider cases such as DMD, where deletions are responsible for 60% of cases.

In other cases such information will not be available. This is inevitable when a gene is first cloned, during the attempt to correlate the presence of mutations with disease symptoms. There are also many genes where a large number of different mutations have been catalogued. Sometimes these mutations are unique to a particular family or individual and are known as **private mutations**. Testing for the presence of mutations in these genes is a formidable technical challenge. Often the genes may be large, and it is possible that the aetiological mutation may be a single base change.

We may thus summarise three different problems that a gene tester will be called upon to solve.

1. The disease is caused by one or a small number of well-characterised mutations. This will be approached by using tests designed to detect specific mutations. These tests are described in section 7.2.
2. The gene is cloned but there are many possible mutations. This requires tests that can detect differences between normal and mutant genes without necessarily defining the nature of the mutation. This is a more difficult challenge requiring a different approach that is described in section 7.2.
3. The mutation cannot be characterised, but the chromosome harbouring the mutation can be tracked through a family. This would have been the normal situation 10 years ago; it is less common now but the need still arises on occasion. An example is given in section 7.2.

Nature of mutations

Before we consider techniques for detecting mutations, it is necessary to consider the different types of mutation that may occur (the nomenclature used to describe mutations is shown in Box 7.1):

- large-scale chromosomal changes such as deletions and inversions, which may be detected by cytogenetic examination;

- small-scale deletions;

- TREs: this type of mutation is easy to identify and proved very useful in the hunt for genes such as the HD gene;

BOX 7.1: NOMENCLATURE USED TO DESCRIBE MUTATIONS

All examples are drawn from mutations in the *CFTR* gene.

Amino acid substitution

Alanine	A	Lysine	K
Arginine	R	Methionine	M
Asparagine	N	Phenylalanine	F
Aspartic acid	D	Proline	P
Cysteine	C	Serine	S
Glutamine	Q	Threonine	T
Glutamic acid	E	Tryptophan	W
Glycine	G	Tyrosine	Y
Histidine	H	Valine	V
Isoleucine	I	Nonsense	X
Leucine	L		

Each amino acid is represented by a single-letter code shown in the table above. The amino acid substitution is represented by a number, describing the residue in the polypeptide, flanked by the single-letter code of the original and replacing amino acid, e.g.

R553X	A truncating mutation in which arginine at position 553 is replaced by a stop codon
N1303K	Asparagine at 1303 is replaced by lysine

Nucleotide substitutions

These are described by a number representing the position of the nucleotide in the cDNA, sometimes prefaced by nt, followed by a letter representing the original nucleotide (G, A, T or C) and an arrow or chevron before the single letter representing the replacing nucleotide. Normally nucleotide changes are specified when they refer to sequences in the intron that affect RNA processing, because nucleotide substitutions in exons that have phenotypic consequences will result in an amino acid change and will be described as such (see above). If they do refer to a nucleotide in an intron, the position of the nearest nucleotide in an exon is given followed by an indication of how many nucleotides before or after the base substitution has occurred, e.g.

621+1 G>T or nt621+1 G→T	The first nucleotide (G) in the intron 3' to nucleotide 621 in the cDNA is replaced by T
1717-1 G>A or nt1717-1 G→A	The last nucleotide (G) in the intron 5' to nucleotide 1717 in the cDNA is replaced by A

Deletions and insertions

Deletions are represented by 'del' or the Greek letter Δ. Insertions are represented by a number, describing the nucleotide before the insertion, followed by 'ins' and a single letter denoting the nucleotide inserted, e.g.

ΔF508	Phenylalanine at position 508 is deleted
394 delT	The nucleotide T at position 394 in the cDNA is deleted
3905 insT	T is inserted after nucleotide 3905

- mutations affecting RNA production and processing, such as promoter or splice site mutations;

- protein-truncating mutations, such as nonsense and frameshift mutations;

- base substitutions that result in a change in the amino acid sequence of the encoded protein.

Some of these mutations, such as protein-truncating mutations or deletions, will quite obviously affect gene function. Others may be more problematic. For example, as discussed in Chapter 9, apparently neutral polymorphisms are common, so it is sometimes difficult to know whether a base change has functional consequences. Sometimes the nature of the change may be significant, for example it may affect a highly conserved amino acid or a splice acceptor site.

Material used for tests

Material used for tests may be recovered from a number of different sources. An important advantage of PCR-based tests is that they can be used with very small amounts of material.

- A single cell from an eight-cell embryo produced by *in vitro* fertilisation; this allows preimplantation testing. Parents known to be at risk of having an affected child produce an embryo by *in vitro* fertilisation. This is then tested for the presence of the genetic disease and only implanted if it is shown not to be affected.

- **Chorionic villous sampling**. The chorion is derived from the zygote but is not part of the developing embryo. It is a membrane with projections or villi that surround the embryo. Chorionic villous sampling is carried out by introducing a catheter through the vagina until it touches the chorion. The test can be safely carried out in the eighth or ninth week of pregnancy. It thus has an advantage over amniocentesis because, combined with PCR amplification of the sample, it allows an earlier termination of pregnancy should this be indicated by the results of the genetic test.

- **Amniocentesis**. Sampling of cells for genetic testing from the amniotic fluid surrounding the developing foetus. The procedure involves inserting a large tube through the mother's abdomen. These cells are cultured *in vitro* for about 3 weeks to increase numbers. The procedure cannot be attempted before the 16th week of pregnancy and is now less commonly used than chorionic villous sampling.

- Blood from the Guthrie card. DNA is stable so the card can be readily posted to different locations or retrieved for tests later.

- Buccal cells recovered from a mouth wash.

- Blood sample, most commonly used for adult tests.

Genetic tests are usually carried out on material amplified by PCR. Both mRNA and DNA can be used as templates. Genomic DNA can be recovered from any convenient source, such as a chorionic villous sample, a blood sample or mouth wash. It contains potentially relevant information in sequences such as promoters that are not expressed in mRNA. The main disadvantage of genomic DNA is that many genes are large with a complex organisation of introns and exons. If the structure of the gene is known, the exons can be amplified by specific primers. DNA samples are usually used to characterise the presence of specific mutations.

Because the introns have been removed by processing, cDNA produced using mRNA as a template is a much simpler target to analyse when the structure of the gene is not known. This is always the case when a gene is first isolated by positional cloning. In this situation it is necessary to prove that the gene is responsible for the condition by showing a correlation between the presence of a mutation and the occurrence of the disease. This requires that the gene from affected and normal individuals is scanned for differences, using techniques described in section 7.3. mRNA is also the starting point for the protein-truncation technique, which requires analysis of the protein encoded by the gene affected and has been found to be superior for use in DNA chips (see below).

7.2 Testing for known mutations

This section considers the general types of test that can be used to detect specific mutations and then illustrates how these are used with examples drawn from the diagnosis of CF, DMD and a TRE mutation.

General principles

Tests for particular mutations are now mostly based on PCR. These tests are designed according to the particular type of mutation being tested (see above). One advantage of PCR is that more than one pair of primers can be included in a single PCR reaction, allowing more than one mutation to be screened in each reaction. Such a reaction is referred to as a **multiplex PCR** reaction. Specific mutations may be recognised by **allele-specific oligonucleotides** (ASOs), which anneal either to a mutant or to wild-type sequence but not both. An important principle with PCR is that tests whose interpretation depends on the presence or absence of an amplification product must have a positive control to ensure that the PCR reaction does not fail for technical reasons, such as an inactive polymerase or impurities in the DNA sample.

General approaches for recognising specific mutations are listed below.

- Large deletions may be recognised by using PCR primers in which either one or both primers anneal with the region deleted. Deleted DNA used as a template fails to produce an amplification band. This test cannot easily be used in heterozygotes, where the wild-type chromosome will produce the expected band. It is possible to carry out quantitative PCR, but as this is technically demanding it cannot be used as a routine procedure. In practice, the use of PCR to detect large deletions is restricted to sex-linked diseases in boys.

- Small deletions may be recognised by amplifying a region of 100 nucleotides that spans the site of the mutation. A deletion is recognised by a band shift in the product. Even differences as small as one nucleotide can be resolved by careful electrophoresis.

- Point mutations, which include both base substitutions and deletions/insertions of only one or two nucleotides, can be recognised by PCR using ASOs as primers. One common form of test based on this system is the **amplification refractory mutation system** (ARMS) test. In this test two PCR reactions are carried out in parallel and the products run in adjacent lanes during electrophoresis. One primer is common to both reactions; the other primer is an ASO that anneals at the site of the mutation. In one reaction the ASO anneals to the wild-type allele, while in the other it anneals to the mutant allele. Thus a band is produced in one or other of the lanes according to whether the template carries the wild-type or mutant allele. Generally, the ARMS test is designed in a multiplex form that allows a number of mutations to be simultaneously screened. This requires careful design of the primers in order to result in well-separated bands upon electrophoresis of the amplification products.

- Sometimes a mutation will either create or destroy a restriction site. This may be simply detected by amplifying a region around the site and attempting to digest the product with the appropriate restriction enzyme.

- TREs can be screened using PCR products that anneal either side of the expansion site. Replication slippage during amplification often results in a ladder effect. This can be very useful for determining the repeat number precisely. Large amplification will sometimes result in the distance between the two primers becoming too large for the PCR reaction to work efficiently. Such cases need to be investigated by Southern blotting.

- ASOs can also be used in hybridisation assays to detect a mutation. Usually the ASOs are attached to the membrane. This is then hybridised to DNA amplified by PCR from the region being tested. A number of different point mutations can be simultaneously screened by having a number of ASOs arranged in an array. DNA chips are an extension of this idea, where the array is miniaturised to allow a very large number of ASOs to be incorporated. In the extreme, an ASO for every single nucleotide position can be included, allowing every possible point mutation to be detected.

Test for mutations in the *CFTR* gene

In people of European extraction, the ΔF508 mutation is the most common CF allele. The proportion of CF alleles that are ΔF508 shows a gradient across Europe, increasing from about 50% in southern European countries to 85% in Denmark and the Basque region of Spain (see Figure 9.8). The average figure is about 70%. Twelve other mutations are also relatively common, having a combined frequency of about 15% (depending on country). The remaining 15% of mutations are individually uncommon, but over 550 have been characterised. The distribution of CF alleles has consequences for genetic testing for CF. Assuming a frequency for the ΔF508 allele of 70% and a combined frequency of 15% for the other common alleles, the frequency of the various classes of CF-affected individuals can be calculated as follows:

49%	ΔF508/ΔF508
21%	ΔF508/common mutation not ΔF508
21%	ΔF508/rare mutation
4.5%	common mutation not ΔF508/rare mutation
2.25%	common mutation not ΔF508/common mutation not ΔF508
2.25%	rare mutation/rare mutation.

These figures will vary in different European countries as allele frequencies change. The important point is that over 90% of CF patients will carry at least one ΔF508 allele. An initial test for the ΔF508 mutation will therefore identify most individuals who are either affected by CF or who are carriers. Testing for ΔF508 together with the 12 common mutations will identify 70% of CF cases in Europeans (more in some countries). However, detecting **compound heterozygotes** between ΔF508 and a rare mutation will be difficult. This is an important complicating factor in embryo testing, neonatal screening or carrier screening in adults. In other populations, different mutations may be more common and testing for *CFTR* mutations must therefore be designed to take account of the genetic structure of the local population or the ethnic origin of the subject.

The ΔF508 mutation may be detected by a simple and robust PCR test that amplifies a 100 bp region spanning the site of the mutation. The 3 bp deletion results in the ΔF508 allele migrating slightly faster than the wild-type allele. As a result, all three possible genotypes may be readily distinguished from each other (Figure 7.1). A multiplex ARMS test can be used to recognise the other 12 most common CF mutations found in European populations. Figure 7.2 shows the result of a multiplex test using a commercial testing kit. This particular test is a slight modification of the original form of the ARMS test. Except for the ΔF508 primers, two parallel reactions are not in the form of pairs of ASOs for mutant and wild-type alleles but contain sets of primer pairs that detect different mutations. External controls were incorporated to ensure the PCR reaction was working.

Sometimes the base change creates or destroys a restriction site, which forms the basis of a very simple test for the presence of the mutation. An example is given in Figure 7.3 for the G551D and R553X mutations, which are the third and sixth most common CF mutations worldwide. Both mutations affect the sequence GTCAAC, which is the target site for the restriction enzyme *Hinc*II. Thus the presence of one or the other is indicated by loss of this restriction site. In addition, the G551D mutation

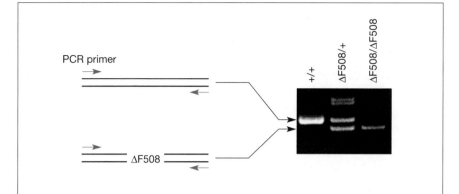

Figure 7.1 A PCR test for the ΔF508 mutation. Chromosomal DNA is amplified using primers that anneal either side of the site of the ΔF508 mutation. The amplification products are separated on a denaturing polyacrylamide gel and visualised by ethidium bromide fluorescence. The wild-type allele is 3 bp larger than the mutant allele so migrates slightly more slowly. The three possible diploid genotypes are clearly distinguishable. The two additional bands in the heterozygote lane are heteroduplexes between wild-type and mutant DNA strands. (Data courtesy of the North Trent Molecular Genetic Laboratory, UK.)

creates an *Mbo*I site, which can therefore be used as a further test to distinguish between the two mutations when loss of the *Hinc*II site is observed.

These and other similar tests provide a simple means for identifying 80–90% of CF mutations. The remaining mutations provide a much sterner challenge because it is not practicable to develop individual tests for each of the hundreds of mutations that have been recorded. Instead it is necessary to use the techniques that scan the gene to detect differences from the wild type without defining the precise nature of these differences. These techniques are reviewed below and some examples of the detection of rare CF mutations are included.

Neonatal screening for CF can be carried out by assaying for levels of immunoreactive trypsin in the Guthrie blood-spot sample. An abnormally high level may indicate CF but it could be due to other causes. Such babies would therefore be subjected to DNA tests. Since most of CF cases would have at least one ΔF508 allele, screening for this mutation would detect nearly all cases. This test can also be carried out on the Guthrie spot blood sample. A diagnosis of CF would be confirmed by continued elevation of immunoreactive trypsin levels after 28 days and by measurement of sweat electrolytes.

Duchenne muscular dystrophy

Two-thirds of DMD cases are caused by large deletions that remove one or more exons. The deletions cluster in two regions: at the 5' end, affecting exons 3–8, and towards the 3' end, affecting exons 44–60 (predominantly 44–50). Multiplex PCR can be used to detect loss of these exons. In this test PCR is used with a mixture of different primer pairs. Each primer pair is designed to amplify a single exon using genomic DNA as a template. The products of the PCR show a band for every exon present; a missing band is

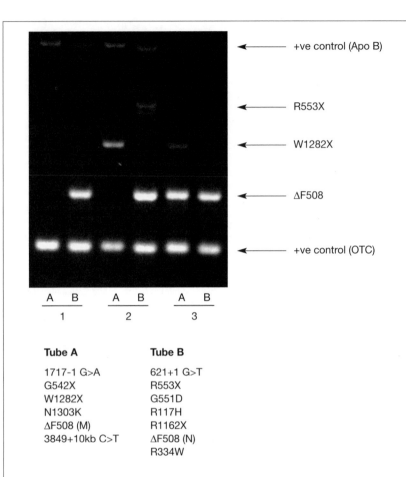

Tube A	Tube B
1717-1 G>A	621+1 G>T
G542X	R553X
W1282X	G551D
N1303K	R117H
ΔF508 (M)	R1162X
3849+10kb C>T	ΔF508 (N)
	R334W

Figure 7.2 Modified ARMS test for *CFTR* mutations. Each test consists of two reactions (lanes A and B), each of which contains six pairs of primers that will only anneal to particular mutant alleles, so a band only appears if the mutation is present. The mutations tested by each tube are listed. In addition, tube B contains a pair of primers that specifically amplifies the wild-type ΔF508(N) allele (note that tube A contains primers that amplify the mutant ΔF508(M) allele). Each lane contains two control reactions to ensure that the PCR reaction is working: one amplifies a portion of the ornithine transcarbamylase (OTC) gene, the other a portion of the apolipoprotein B (Apo B) gene. The Apo B reaction is deliberately designed to be suboptimal so that it fails first if there is a problem with the PCR. The figure shows the result of tests on three individuals. The PCR products have been separated in an agarose gel and visualised by ethidium bromide fluorescence. Individual 1 does not contain any of the tested mutations, as there is no amplification product in lane A and there is a band corresponding to the wild-type ΔF508(N) allele in lane B. Individual 2 is a W1282X/R553X compound heterozygote, as there is a band corresponding to W1282X in lane A and a band corresponding to R553X in lane B. Individual 3 is a W1282X/ΔF508(M) compound heterozygote, as there are two bands in lane A corresponding to W1282X and ΔF508(M). Although the Apo B controls in the two lanes from individual 3 were very weak, the presence of the other bands confirmed that the PCR reaction was working normally. (Data courtesy of the North Trent Molecular Genetic Laboratory, UK.)

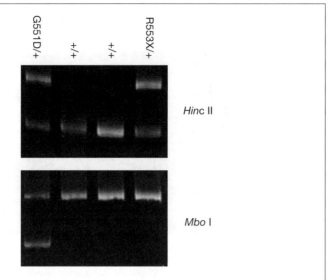

Figure 7.3 Diagnostic restriction enzyme assays for the *CFTR* G551D and R553X mutations.
The wild-type sequence, G̲G̲T̲C̲A̲A̲C̲ is a target for *Hinc*II (underlined). This is altered to G̲A̲T̲C̲A̲A̲C
in G551D, which contains a target for *Mbo*I (underlined) but is no longer digested by *Hinc*II. In
R553X the wild-type sequence is altered to GGTCAAC, which is not a target for either enzyme. In
the test, a fragment containing the mutation site is amplified by PCR and digested separately
with each enzyme. If the individual contains either mutation, the *Hinc*II digest fails and so a
heterozygote produces two bands (the wild-type allele is still digested). When digested with
*Mbo*I only a G551D heterozygote produces two bands because one allele is a target for this
enzyme. (Data courtesy of the North Trent Molecular Genetic Laboratory, UK.)

therefore diagnostic of a deletion. Figure 7.4 shows an example of this multi-plex PCR test used to screen for deletions of exons in the 3' deletion hotspot. As discussed above, this test cannot be readily used to detect carrier females because the exons from the wild-type chromosome will be amplified.

The remaining cases of DMD not caused by deletions are mostly non-sense or frameshift mutations that cause chain termination; these may be detected by the protein-truncation test described below.

Trinucleotide repeat expansions
Each TRE is a separate event; however each mutation affects the same site and so can be readily detected by a PCR reaction based on primers that anneal either side of the expansion site. The size of the PCR product indicates whether an expansion event has occurred. The extent of the expansion may be important in the prognosis of age of onset and severity; this information is also revealed by the expansion. Figure 7.5 shows an example, where spinocerebellar ataxia type 2 is diagnosed by such a test.

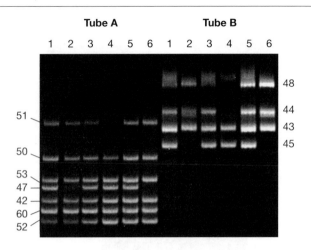

Figure 7.4 Multiplex test for deletions in the dystrophin gene. Each test consists of two tubes (A and B) containing pairs of PCR primers that amplify specific exons of the dystrophin gene, as shown at the sides. Individual number 6 has a deletion of exons 45 and 47, as judged by the missing bands in tubes B and A respectively. This probably results from a single deletion that spans exon 46, which was not tested. Individual 2 has a deletion of exon 45; the band for exon 47 is weak in this individual but was clearly visible in the original photograph. Individual number 4 apparently has a deletion of exons 44, 48 and 51. These exons are non-contiguous and so this result would be regarded with suspicion and the sample retested. The most likely explanation is that the PCR reaction failed to amplify the three largest bands in this sample. (Data courtesy of the North Trent Molecular Genetic Laboratory, UK.)

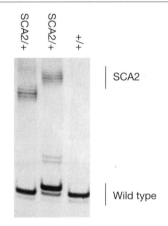

Figure 7.5 TREs in spinocerebellar ataxia type 2 (SCA2). The CAG repeat sequence was amplified using PCR primers that anneal to flanking regions. SCA2 heterozygotes have an extra band corresponding to the amplified allele. The samples have been separated on a polyacrylamide gel and visualised by silver staining. (Data courtesy of the North Trent Molecular Genetic Laboratory, UK.)

Section 7.2 considered some case studies where the nature of the likely mutation is known and it is possible to design specific tests to identify them. However, many genes are affected by a large number of different mutations; 10–20% of CF alleles are rare mutations. Genes such as *BRCA1* and *BRCA2* are affected by a large number of different mutations. Short of sequencing each gene being tested, it is necessary to use techniques that can scan a sample and detect differences from the wild-type gene. One ever-present problem in this type of analysis is deciding whether a change detected is a 'polymorphism' or a 'mutation', i.e. whether the change has deleterious consequences for gene function.

Single-stranded conformational polymorphisms

Single-stranded nucleic acid molecules form extensive secondary structure in solution due to intramolecular base pairing. This secondary structure has a large effect on the rate of migration during electrophoresis under non-denaturing conditions. Even a single base change can radically alter the secondary structure and hence the rate of migration of a nucleic acid fragment. This forms the basis of the single-stranded conformational polymorphism (SSCP) technique, which allows the presence of a base change to be detected without defining the nucleotide affected. PCR is used to amplify the gene in short sections of under 200 bp and the products are separated under non-denaturing conditions by polyacrylamide gel electrophoresis. An example is shown in Figure 7.6 where a rare mutation in the *CFTR* gene was detected. This technique is very simple and cheap and detects 70–95% of base changes in fragments under 200 bp.

Denaturing gradient gel electrophoresis

Partially denatured DNA molecules migrate at a much slower rate during gel electrophoresis. If a molecule is subjected to electrophoresis in a gradient of denaturing agents, such as heat or chemical denaturants, it will migrate to the point where it starts to denature and then effectively stop migrating. This is called denaturing gradient gel electrophoresis (DGGE). The point at which denaturation starts is extremely sensitive to sequence and will be altered by even a single base change. Mutations may therefore be detected by a bandshift in PCR products from the test DNA compared with wild type (Figure 7.7). This method detects about 99% of mutations and is simple to operate once the conditions are fully established for each DNA fragment tested. However, designing the right conditions is not easy, and the conditions for each fragment have to be developed separately.

Mismatch cleavage detection

If a wild-type DNA molecule is hybridised to a mutant sequence, at the site of the mutation there will be a mismatch in the double-stranded duplex. This mismatch may be cleaved chemically using agents such as piperidine or osmium tetroxide (chemical mismatch cleavage, CMC) or by enzymes

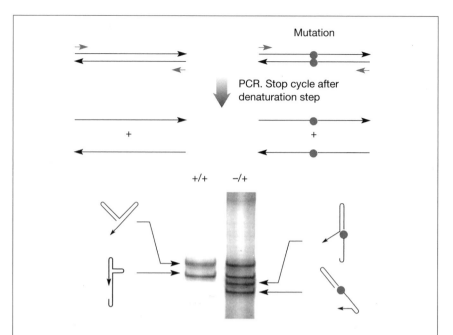

Figure 7.6 Detection of a rare *CFTR* mutation using SSCP. Reference wild-type DNA and test DNA samples are amplified by PCR and the PCR reaction stopped at the denaturation stage of the cycle. The products are loaded on to a native polyacrylamide gel that allows the single-stranded molecules to form secondary structure, which affects their rate of migration. Since each strand takes up a different structure, two bands are seen in wild-type DNA. Even a single base change alters the secondary structure, so a mutant DNA molecule produces two bands in a different position and a heterozygote will produce four bands (two wild type, two mutant). (Data courtesy of the North Trent Molecular Genetic Laboratory, UK.)

involved in mismatch repair (enzyme mismatch cleavage, EMC). DNA is amplified from wild type and DNA to be tested. The products are separately denatured and then allowed to form heteroduplexes, before being subjected to CMC or EMC. The products are fractionated by DGGE; a mutant DNA strand is detected by the appearance of a smaller fragment (Figure 7.8). The use of chemical cleavage agents is problematic because they are highly toxic and carcinogenic. Enzymic cleavage is now well developed and is claimed to detect 99% of mutations. However, as yet it is not widely used.

Protein truncation test
In many genes the majority of disease-causing mutations result in truncation of the protein product. The presence of such a mutation may be detected by using the product of a PCR reaction as a template for an *in vitro* transcription/translation reaction. The truncated protein from a mutant gene is detected by gel electrophoresis. The advantage of this test compared with those described above is that it will only detect changes that are biologically significant. The problem with the other methods is that many biologically neutral polymorphisms will be detected and there is no

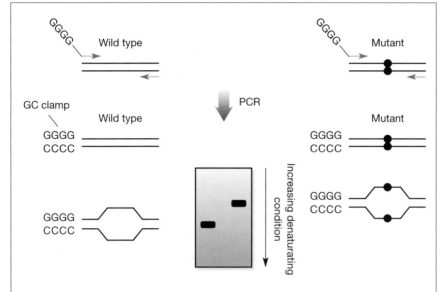

Figure 7.7 An example of DGGE. Reference wild-type and test DNA are amplified by PCR. The 5' end of the upstream primer is extended with a series of C residues, resulting in a 'GC clamp' in the amplified product. The products are separated in a polyacrylamide gel with a gradient of increasingly denaturing conditions. When the duplex DNA molecules start the migration, their motion is dramatically slowed. This point is sensitive to a base-pair change in sequence, so mutant and wild-type DNA molecules migrate to different positions. The GC clamp is slow to denature because of the three hydrogen bonds in a GC base pair. This stabilises one end of the molecule and has been found empirically to increase the discrimination between mutant and wild-type DNA molecules.

simple way of distinguishing these from disease-causing mutations. This test is likely to be the method of choice for screening for mutations in tumour-suppressors such as *BRCA1*, *BRCA2*, and *APC* genes, where over 90–95% of mutations are chain terminating. An example of its use in detecting a mutation in *BRCA2* is shown in Figure 7.9.

DNA chips
DNA chips are a novel and still developing technology that allow the identity of every nucleotide in a test DNA sequence to be examined in a single operation. The principle is to attach a massive array of oligonucleotides on to a solid glass or silica support. For each position in the DNA being tested there is a separate array of oligonucleotides. The sequence of each oligonucleotide is the same as the wild type, except that the central base is systematically changed so that there is an oligonucleotide for every possible sequence variation. The test DNA is labelled with a fluorescent dye and hybridised to the chip. A scanning confocal fluorescence microscope is used to record the result. The test DNA preferentially hybridises to the oligonucleotide in each set that exactly matches its sequence, thus indicating the identity of the nucleotide at each position.

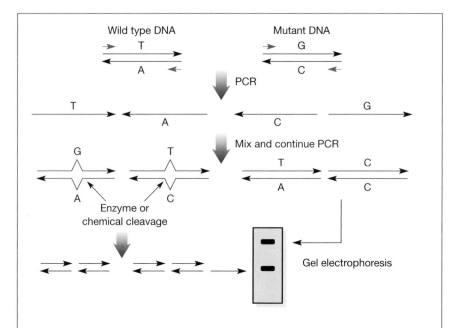

Figure 7.8 Mismatch cleavage detection. Reference wild-type and test DNA samples are amplified separately by PCR for a number of cycles, then mixed and subjected to further rounds of PCR amplification, which allow heteroduplexes to form if a mutation is present in the test DNA. Chemicals (such as osmium tetroxide or piperidine) or mismatch repair enzymes are used to cleave the DNA at the site of any mismatches. Cleaved molecules are smaller and therefore migrate faster on the gel.

Recently a chip was constructed that recognises every base in the 3.45-kb exon 11 of the *BRCA1* gene. The way this DNA chip works is described in Figure 7.10. A key technology in the construction of a DNA chip is photolithography, used to make integrated electronic circuits. This was used to synthesise the oligonucleotides directly on the surface of the chip. This technology allowed the 90 000 oligonucleotides used in the BRCA1 chip to be attached to an array that measured 1.28 × 1.28 cm.

DNA chips can be designed to recognise specific mutations rather than examine every base pair. This may make interpretation of the results simpler since it would only identify mutations that are definitely known to cause disease, rather than identifying all base changes, many of which may be neutral polymorphisms. Chips have already been constructed that screen for known mutations in the *CFTR* gene, the β-globin gene and the mitochondrial genome.

DNA sequencing
Sequencing the test DNA is the most direct way of determining whether a mutation is present. As stated above, re-sequencing a large gene for a rou-

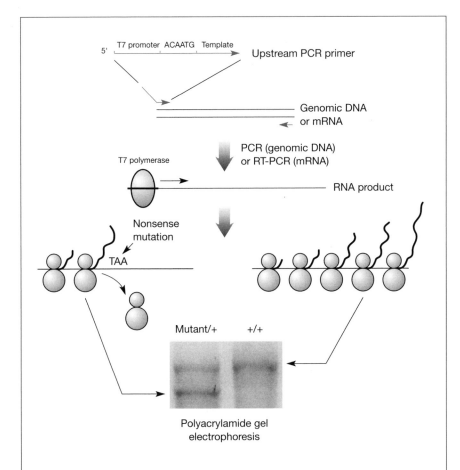

Figure 7.9 Detection of a *BRCA2* mutation by the protein truncation test. A section of the *BRCA2* gene was amplified by PCR. The design of the upstream primer is important. It has a 5' extension that contains the promoter sequence for the phage T7 RNA polymerase followed by a consensus eukaryote protein translation initiation sequence. The primer is designed so that the reading frame of the amplified *BRCA2* sequence will be correctly aligned with the ATG of the translation initiation sequence. The amplified product can be transcribed *in vitro* using phage T7 polymerase and the resulting RNA translated *in vitro* using ^{35}S-labelled methionine to visualise the product by autoradiography. If the *BRCA2* sequence contains a nonsense mutation, translation is halted prematurely and the product is truncated. This results in a faster rate of migration in the polyacrylamide gel used to fractionate the products of the *in vitro* translation. (Data courtesy of the North Trent Molecular Genetic Laboratory, UK.)

tine test is not a practicable proposition. However, as we shall see in the next chapter, improvements in sequencing efficiency are necessary for the completion of the Human Genome Project. Already the Sanger dideoxy sequencing method has been automated and is very much more efficient than it was only a few years ago. It is possible therefore that sequencing genes to screen for mutations may become routine in the next few years.

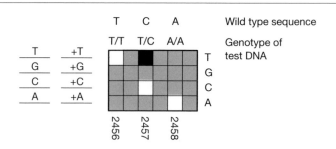

Figure 7.10 Detection of a heterozygous base substitution in the *BRCA1* gene using a DNA chip. The left-hand part of the diagram shows the organisation of an oligonucleotide array that interrogates a single base position in the test DNA. Each oligonucleotide is 20 nucleotides long; the sequence matches the wild-type sequence except that there is a substitution or insertion at residue 11 as shown. Wild-type and test RNA sequences were fluorescently labelled with different coloured dyes and then hybridised separately to the chip. The composite image for three consecutive positions is shown on the right. The oligonucleotides are attached to the chip in square cells. Cells shown in white hybridised to wild-type and test sequences with equal intensities, blue-coloured cells hybridised to neither target and the black cell hybridised to the test RNA but not to the wild-type RNA. The test RNA sequence hybridised to the cells containing T, C and A in the oligonucleotides corresponding to the wild-type sequence, but it also hybridised to the cell that contained T at position 2457 showing that one allele contained a T at this position. The individual from which the test RNA was derived is therefore heterozygous for a base substitution at position 2457. Quantitative analysis of the hybridisation signal also detected the mutation, as the wild-type DNA produced a stronger signal than the test DNA at position 2457C. RNA was used because it was found to give superior results compared with DNA. The RNA was generated from PCR-amplified DNA by *in vitro* transcription. As well as the oligonucleotides shown, oligonucleotides with various deletions were also included in each array.

Chromosome tracking

When the mutation causing a genetic disease segregating in a family cannot be identified, it is necessary to track the chromosome carrying the mutation in order to diagnose the status of an embryo. This involves genotyping linked polymorphic markers and establishing the phase of markers and mutation. Highly polymorphic microsatellites are normally used. An example is shown in Figure 7.11.

7.4 Summary

- Gene testing has been revolutionised by the isolation of disease genes and the amplification of small amounts of material by PCR.

- Gene testing is carried out in a number of different situations:
  - To diagnose embryos known to be at risk of genetic disease.
  - Neonatal screening.
  - As an aid to medical diagnosis.
  - To diagnose the status of an adult at risk of a late-onset disease.

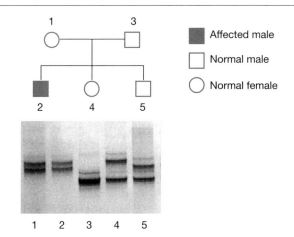

Figure 7.11 Tracking Prader–Willi syndrome in a family. Prader–Willi syndrome (PWS) is characterised by short stature, obesity and mental retardation. It is normally caused by a deletion on the paternally derived copy of chromosome 15. The corresponding region in the maternally derived chromosome is inactivated by a process known as genomic imprinting during oogenesis. In the family shown here, a closely linked microsatellite polymorphism has been tracked using PCR. The proband (individual 2) has apparently inherited both copies of chromosome 15 from his mother. The explanation of non-paternity was ruled out by other markers not shown here. Because both maternal copies of the PWS locus are inactivated by genomic imprinting, there is no functional protein and PWS results. Based on this analysis, the family can be assured that the father does not carry the deletion and if the mother becomes pregnant again a re-occurrence of the disease is most unlikely. The inheritance of two different copies of a chromosome from the same parent is known as **uniparental heterodisomy.** Exactly how it occurs is not fully understood, but it is thought that non-disjunction results in an egg with two copies of chromosome 15. The resulting zygote will be trisomic, but compensation can occur resulting in the loss of one copy of chromosome 15. If the paternal chromosome is lost, then both remaining copies will be maternal in origin. (Data courtesy of the North Trent Molecular Genetic Laboratory, UK.)

- ○ To determine susceptibility to a complex disease.
- ○ To confirm the identity of a candidate gene by showing a correlation between disease and mutation.

- Some mutations, such as protein-truncating mutations, will clearly ablate protein function. Others, such as base substitutions, may be neutral polymorphisms.

- DNA or RNA may be used for the tests. DNA is easy to obtain, but the extremely large size and complex organisation of some genes may make searching for unknown mutations difficult. RNA from the right tissue may be more difficult to obtain but it will represent the gene sequence in a more manageable form.

- PCR tests have been developed to recognise the different types of mutations, e.g. large deletions, small deletions and point mutations. One commonly used test is the ARMS test, which uses ASOs as primers to test for the presence of several specific mutations in two parallel PCR assays.

- The *CFTR* ΔF508 mutation may be detected by a simple PCR test. A multiplex PCR test can be used to detect other common mutations.

- About 60% of DMD cases are caused by deletions in the dystrophin gene. These may be detected by a multiplex PCR test.

- TREs can be detected by amplifying the expansion site using PCR primers that anneal to flanking regions.

- A general method for detecting known mutations is the use of ASOs that only hybridise to DNA containing a particular mutation.

- Unknown mutations may be detected by a number of tests that attempt to find differences between the test sample and a reference wild-type sequence.

- SSCP relies on the fact that the rate of migration of single-stranded DNA through a native gel is highly sensitive to secondary structure, which in turn depends on sequence.

- DGGE is based on the difference in migration rate of duplex and partially denatured DNA. The point at which a DNA molecule starts to denature is dependent on its exact sequence.

- Mismatch cleavage uses chemicals or enzymes to cleave a heteroduplex between test and reference DNA at the site of a sequence mismatch.

- The protein truncation test is based on *in vitro* transcription/translation. If the reference sequence contains a chain-terminating mutation the polypeptide produced will be smaller than the reference.

- DNA chips are, in principle, capable of detecting all possible changes in a gene sequence in a single operation. They are still at an experimental stage.

- Direct DNA sequencing for routine testing is not currently practicable, but it may become easier with improved sequencing technology.

- When a family requires genetic testing and counselling for a genetic disease in which the aetiological mutation has not been identified, it is necessary to track the chromosome carrying the mutation. This is now normally done with linked microsatellite markers amplified by PCR.

Further reading

General
ENG, C. AND VIJG, J. (1997) Genetic testing: the problems and the promise. *Nature Biotechnology*, **15**, 422–426.

Cystic fibrosis
FERRIE, R.M., MARTIN, M.J., ROBERTSON, N.H. *et al.* (1992) Development, multiplexing, and application of ARMS tests for common mutations in the CFTR gene. *American Journal of Human Genetics*, **51**, 251–262.

Methods for unknown mutations
GROMPE, M. (1993) The rapid detection of unknown mutations in nucleic acids. *Nature Genetics*, **5**, 111–116.

JONSSON, J.J. AND WEISSMAN, M.S. (1995) From mutation mapping to phenotype cloning. *Proceedings of the National Academy of Sciences USA*, **92**, 83–85.

MASHAL, R.D., KOONTZ, J. AND SKLAR, J. (1995) Detection of mutations by cleavage of DNA heteroduplexes with bacteriophage resolvases. *Nature Genetics*, **9**, 177–183.

YOUIL, R., KEMPER, B.W. AND COTTON, R.G.H. (1995) Screening for mutations by enzyme mismatch cleavage with T7 endonuclease VII. *Proceedings of the National Academy of Sciences USA*, **92**, 87–91.

Detection of *BRCA1* mutations using a DNA chip
HARCIA, J.G. BRODY, L.C., CHEE, M.S., FODOR, S.P.A. AND COLLINS, F.S. (1996) Detection of heterozygous mutations in *BRCA*1 using high density oligonucleotide arrays and two-colour fluorescence analysis. *Nature Genetics*, **14**, 441–447.

CHAPTER EIGHT

The Human Genome Project

Look on internet

Key topics

- Rationale for the Human Genome Project
- History and goals of the Human Genome Project
- Sequencing the human genome
  - Strategies for obtaining sequence-ready clones
  - Sequencing technology
  - Sequencing centres
- Informatics
- Model organisms
  - *Saccharomyces cerevisiae*: the first eukaryote genome to be sequenced

8.1 Introduction

The **Human Genome Project** is an international collaboration to map and sequence the entire 3000 Mb of DNA that make up the human genome. The fundamental rationale for such an immense undertaking is that if we wish to understand how the human organism is constructed and functions at a molecular level then there is no better place to start than examining the blueprint.

The nucleotide sequence contains the information that specifies the amino acid sequence of every protein and functional RNA molecule. In principle it will be possible to identify every protein responsible for the structure and function of the human body. The pattern of expression in different cell types will specify where and when each protein is used. The amino acid sequence of the proteins encoded by each gene will be derived by the conceptual translation of the nucleotide sequence. Comparison of these sequences with those of known proteins, whose sequences are stored in databases, will suggest an approximate function for many proteins.

Moreover, as our knowledge of protein structure increases it is hoped that computer algorithms can be developed that will allow structure to be predicted from amino acid sequence.

It is estimated that there are 65 000–80 000 such proteins (see Chapter 2), with the identity of about 2000 known at present. Understanding the role of each protein will undoubtedly occupy biologists for many decades to come. This task will be aided by the study of the same components in model organisms where more sophisticated experimentation can be used to study their function. Sequencing the genomes of model organisms is therefore also part of the Human Genome Project.

As well as aiding the understanding of the function of human genes, the model organisms being studied span the range of biological diversity from bacteria to mice. The genomic sequences of representative organisms such as these will illuminate how each is constructed and functions, so the project is relevant to the whole of biological research. Moreover, there will undoubtedly be biotechnological benefits, not only from the general understanding of biological principles but also from the development of technology to rapidly sequence the genomes of economically important organisms.

The genome is more than the sum of its coding sequences. We know other sequences are important. Proper control of gene expression is often as important as the information content of the genes themselves. Sequences upstream of genes control the pattern of their transcription, while other sequences control RNA splicing. Some DNA sequences are essential for the maintenance of chromosome integrity: telomeres protect the ends of chromosomes and centromeres are required for the segregation of chromosomes in meiosis and mitosis (see Chapter 2). We suspect that the majority of DNA has no function or, if it does, a function that does not depend on its sequence. Sequencing non-coding as well as coding regions will contribute towards an understanding of chromosome structure. It will inform us about the role of non-coding sequences that are clearly functional. It will help resolve the question of whether the remaining DNA does have a function and, if it does not, how it evolved and why its presence is tolerated.

Human diversity

Susceptibility to many diseases is genetically determined, at least in part. While the exact aetiological mutations can be described in most monogenic diseases, they are far less easy to specify in the more common complex diseases. Knowing the entire sequence of the human genome will allow the exact differences between susceptible and non-susceptible individuals to be determined. This will identify the proteins involved and allow their normal function to be studied. Ultimately, it may be possible to design drugs to correct, or compensate for, the dysfunction of the protein in the disease state. Moreover, knowing the exact nature of the proteins involved will allow the effect of the treatment to be monitored at a molecular level and adjustments made accordingly.

As well as variation between individuals within a single population, there is systematic variation between populations. This variation allows the recent evolution of humans to be studied (see Chapter 9). Much of the present work is based on either nuclear polymorphisms analysed at the protein level or mitochondrial sequence variation. The ability to compare whole genomes of individuals in different populations will enormously enhance the study of human population genetics. There is an important practical aspect to this. Certain populations have much lower rates of certain diseases (see Table 9.1). These differences are, at least in part, genetically determined. Understanding the basis of this will contribute to the understanding and treatment of these diseases. There is some urgency to this side of the project. We shall see in Chapter 9 that the differences between populations were probably established when population numbers were small during the **palaeolithic** period. During the last 500 years, world populations have been mixed by large-scale migrations. There are some **aboriginal populations** that may soon become extinct, but whose gene pool may prove a valuable subject for study.

Sequencing the entire genome and identifying every protein will be very difficult

Sequencing the human genome will be a formidable task and it may not be possible to achieve all the idealised objectives outlined above. The way the information is structured in the human genome will make it difficult to derive its sequence and then to interpret it. Large tracts of DNA, for example centromeric heterochromatin, are simple tandem repetitive sequences that can extend for megabases. These regions are technically extremely difficult to clone and sequence: it may not be sensible or even possible to try.

The fraction of the genome that encodes proteins is very low, probably of the order of 3%. Some genes are extremely large: the dystrophin gene extends over 2.4 Mb of the X chromosome; there are 79 different exons, the average size of which is 177 bp, and the average size of each intron is 30 kb. Would it have been possible to reconstruct the amino acid sequence of the dystrophin protein purely from the nucleotide sequence of the genomic locus? There are tools, apart from pure sequence analysis, to identify expressed regions (see Chapters 3 and 4). Nevertheless, identifying every gene is going to be a formidable challenge. It may never be possible to be completely certain that the identity of every protein is known.

8.2 History and goals of the Human Genome Project

8.2.1 Inception of the Human Genome Project

The Human Genome Project was initiated in the mid-1980s. It originated from two independent sources. Firstly, the US Department of Energy (DOE) had a mandate to monitor the effects of low levels of environmental radiation and chemical mutagenesis. Two geneticists, Ray White and Mortimer Mendelsohn, convened a meeting in Ulta, Utah, to consider how this brief could be fulfilled. The conclusion was that it was necessary to be able to detect very low frequencies of nucleic acid changes and the only way this could be done was to sequence the genome to provide a reference sequence. In 1986, DOE announced its Human Genome Initiative after a meeting in Santa Fe that considered its feasibility.

The other strand was the community of biomedical scientists, who were just starting to come to grips with the problems of cloning human genes outlined in Chapter 4. At two meetings, in Santa Cruz in 1985 organised by Robert Sinsheimer and at Cold Spring Harbor Laboratory in 1986, the idea of a collaborative project emerged. The DOE designated particular sequencing centres at Lawrence Livermore Laboratory, the Los Alamos Laboratory and the Lawrence Berkeley Laboratory. The major source of funding for biomedical science in the USA is the National Institutes of Health (NIH). NIH established a special office called the National Centre for Human Genomic Research (NCHGR), whose first director was the Nobel Laureate James Watson and subsequently Francis Collins. In 1988 scientists involved in the project established an organisation called the Human Genome Organisation (HUGO) to coordinate international collaboration.

The US Congress formally agreed funding in 1991 to complete the project at an estimated total cost of $3 billion. In the USA these funds are administered by the DOE at its own laboratories and the NIH through peer-reviewed projects coordinated by NCHGR. Other countries have funded their own contributions to the project through public and charitable funds. Recently the Wellcome Trust in the UK has provided £50 million to improve the sequencing capability of the Sanger Centre (see Box 8.1). In France the Muscular Dystrophy Association has provided funds for genome mapping at the Genethon laboratory. As well as public funds and finance from charities, private funds have been important for both commercial genome research and support of research in academic institutions. For example, the pharmaceutical company Merck funded the construction of an EST database at Washington University, St Louis and a privately financed organisation called Human Genome Sciences (HGS) has also made a substantial investment in EST collection and sequencing. The interaction between privately financed and public research has not always proceeded smoothly. Major controversies have arisen over the conditions that HGS initially imposed over access to its databases.

At first, the idea of the Human Genome Project was opposed by many who considered it a form of 'stamp collecting', i.e. gathering information for its own sake instead of carrying out experiments to test hypotheses. One of the problems was perceived to be the size of the task, which made the work seem both excessively uninteresting to carry out and so expensive that it would drain funds from other forms of research. It was appreciated from the beginning that before work on the project itself could commence it was necessary to improve the technology. The development of the STS mapping marker (see Chapter 3) is an example of the effect that a methodological innovation can have.

It was also appreciated that the medical advances that would arise could have a negative as well as a positive impact on society. It was therefore necessary to investigate legal, ethical and social issues alongside scientific developments. These issues are considered in Chapter 11.

8.2.2 Five-year goals for the Human Genome Project

In 1988 the DOE and NIH jointly agreed a strategic plan for work on the project in the USA. It was to start in 1990 and was projected to take 15 years. Specific goals were set for the first 5 years, which were seen to be a necessary preliminary before the task of sequencing could begin. These 5-year goals were revised in 1993 to take the project to 1998.

Goal 1: A genetic map with a resolution of 2–5 cM

Chapter 3 discussed the crucial importance of genetic maps. They make the connection between biological and medical phenotypes and the underlying genome. The original goal was a genetic map with a resolution of 2–5 cM. The completion of the genetic map in 1995 was the first of the goals to be achieved. The average distance between markers was 1.6 cM, which exceeds original specification. As well as the development of the map itself, this goal incorporates the development of markers that are easy to use and methods for genotyping that can be readily automated. The genome scanning techniques discussed in Chapter 5 illustrate both the reason for this goal and the success that has been achieved in attaining it.

Goal 2: A physical map of the human genome

Physical maps are necessary for two reasons: firstly, to locate the actual DNA sequences that form genes and, secondly, to order clones prior to sequencing. The goal is an STS content map (see Chapter 3) with an STS site every 100 kb. This will require the mapping of about 30 000 STS markers. A map of 15 000 markers was published in 1996 and the full map will probably be available by the end of 1997. There will be a continuing need to develop chromosome maps to a higher resolution to provide sequence-ready clones. This topic is discussed in more detail below.

Goal 3: Develop DNA sequencing technologies

The second phase of the Human Genome Project is the determination of the nucleotide sequence of the human genome. Before this can begin in earnest, it is necessary both to develop an effective strategic approach to prepare the clones for sequencing and improve the technology used to derive the sequence of an insert in a suitable clone. We discuss both of these issues in a separate section below. The formal goals for sequencing in this phase of the project were:

1. develop efficient approaches to sequencing DNA regions several megabases in size;
2. develop technology for high-throughput sequencing;
3. build up the capacity to collect sequence at a rate of 50 Mb per year by the end of 1998.

Goal 4: Gene identification

Although not part of the original project, the ability to locate genes in physical and genetic maps and in the nucleotide sequence as it becomes available is an absolutely essential requirement to make the data intelligible and useful. Chapter 3 considered how expression maps based on a special form of STS, called an EST, can be aligned with the physical and genetic maps in order to provide an integrated resource to find genes. The development of such maps was incorporated into the formal goals of the project in 1993.

Goal 5: Technology development

Chapter 3 described how robotics and automatic data collection were essential to allow the construction of detailed maps of the genome. This technology will be equally essential to derive the nucleotide sequence. It is wasteful if each centre develops its own technology separately, instead of pooling their expertise. The Human Genome Project therefore aims to encourage the development of improved technology.

Goal 6: Model organisms

Sequencing of model organisms is important for two reasons. Firstly, the simpler organisms provide an opportunity to refine strategies and technologies that will be used on the more complex human genome. Secondly, many genes show evolutionary conservation across a diverse range of species. In many cases it will be easier to understand their function in a simpler organism that is more amenable to experimental manipulation. This approach is based on the concept that the fundamental processes of life are conserved in all organisms. The cellular organisation of a simple eukaryote such as a yeast is remarkably similar to that of a more complex organism such as a mammal. The workings of the cell cycle engine, signal transduction mechanisms, the cytoskeleton and secretory pathway, are fundamentally similar in all eukaryotes. We discuss this point in more detail in a later section.

In metazoan organisms, the developmental circuits that programme development are similarly extensively conserved. It is often much easier to study these processes in a laboratory model than in humans. Genes whose sequence similarity in two different species suggests a common evolutionary origin are said to be **homologous**. Often some parts of homologous genes are highly conserved while other parts have diverged. This helps identify functionally important parts of the protein. A number of genome projects are currently underway or have been completed. Some of these are listed below. The figure in parentheses refers to the genome size.

- *Escherichia coli* (4.6 Mb). Historically, the laboratory organism used to investigate many fundamental processes. It is also the workhorse for molecular genetic techniques such as gene cloning. The complete sequence of this organism was announced in 1997.

- *Saccharomyces cerevisiae* (12.068 Mb). The budding yeast used for the fermentation of alcoholic beverages and bread-making. This organism has been intensively studied as a model eukaryote. The complete sequence of this organism was announced in 1996. The importance of this achievement and the lessons that can be derived from it are discussed below.

homologus.

- *Caenorhabditis elegans* (100 Mb). This nematode or roundworm has been chosen as a model organism for developmental studies because it consists of only 959 somatic cells. The lineages of each one of these cells from the fertilised zygote and the connections made by every single neurone have been documented. The aim is to decipher the genetic control circuits operating in order to provide a complete description of the development of a metazoan organism. The complete sequence is expected to be available by the end of 1998.

- *Drosophila melanogaster* (165 Mb). Historically, the fruit fly is the organism in which many of the basic principles of genetics were elaborated. Its current importance lies in the way that analysis of developmental mutants led to the discovery of the homeobox family of genes that control development in all metazoan organisms.

- *Arabidopsis thalania*. A member of the mustard family of plants. It has the smallest genome and gene density so far identified in flowering plants. It has therefore been selected as the representative genome of higher plants. The project was initiated in September 1996 and is projected to take 3 years. The information will be of importance to all economically important plants.

- *Fugu rubripes rubripes* (400 Mb). The puffer fish, a vertebrate with an unusually small genome. It contains far less repetitive DNA and introns are smaller, yet its coding sequences show high sequence conservation with other vertebrates.

- *Mus musculus* (3000 Mb). The mouse is used extensively as a model mammal in biomedical research. Sophisticated molecular technology allows the construction of **knockout mice**, animals in which specific genes have been deleted, or **transplacements**, where genes have been replaced with genes carrying specific mutations. This has provided models of various human monogenic diseases and allowed the role of proteins such as p53 to be investigated. Meiotic genetic analysis has allowed susceptibility to complex diseases and disorders such as diabetes and obesity to be identified. **Synteny** (conservation of gene order) will often lead to the human homologue of a medically important gene identified by genetic analysis in mice. The sequence of the mouse genome will be similar to the human genome, but will afford opportunities for experimental investigation not available in humans. A detailed genetic map of the mouse genome has already been constructed alongside the human genetic map.

Goal 7: Informatics

The success of the Human Genome Project is heavily dependent on the way that the information is collected, organised and interpreted. This places heavy demands on information technology. The list below gives some idea of the areas where this is important.

- Construction and use of physical and genetic maps.

- Assembly of sequencing data from individual sequencing reactions.

- Recognition of protein-coding sequences and other features of interest, such as enhancers, transcription start sites, splice acceptor sites, etc.

- Comparison of amino acid sequences with other genes to detect homologies and recognise functional motifs. This requires the maintenance of computer databases and the efficient means to search for protein sequences similar to a test sequence.

- Prediction of protein structure from amino acid sequence. This has been a long-term target of protein structure research for many years and has proved very difficult to achieve. However, the number of proteins whose structure is known is rapidly increasing. In turn, we may expect the algorithms that predict structure from amino acid sequence to improve.

- Analysis of the long-range sequence organisation of chromosomes in order to recognise features important in their function and evolution.

Goal 8: Ethical, legal and social issues
The relationship between human molecular genetics and society is examined in Chapter 11. In the USA, the Human Genome Project funds research in these areas, demonstrating its recognition of their importance. This goal also incorporates the need to foster public and professional education concerning human molecular genetics.

Goal 9: Training
The need for training in relevant disciplines has led to formal training programmes in many countries as well as international schemes, such as those funded by the European Community and the European Molecular Biology Laboratory (EMBL).

Goal 10: Technology transfer
Genome research is being carried out in a variety of contexts: large genome centres, smaller publicly funded laboratories and commercial laboratories. There is therefore a need to encourage transfer of technologies between these various centres. Furthermore, many important instruments used in the research have been developed and marketed by private organisations. An example of this is the DNA sequencer manufactured by Applied Biosystems, which has become the standard instrument for DNA sequencing and for genotyping microsatellites using fluorescence-based PCR assays in genome scanning.

Goal 11: Outreach
The Human Genome Project fosters cooperation among genome centres and individual laboratories for the rapid release of data and distribution of

clones and other genome reagents. A novel aspect of the project is the immediate distribution of data as they are generated rather than the normal process of scientific publication, which can result in considerable delay between the time the data are collected and when they become publicly available. The importance of rapid publication is that those funded to collect the sequence information should not be seen as having prioritised access to it for their own in-house research projects. A condition of NIH- and DOE-funded research is that data are released within 6 months of their collection. The Sanger Centre in the UK releases its data daily on to the World Wide Web.

8.3 Sequencing the human genome

The major goals set for mapping the human genome have now been, or soon will be, completed. The remaining task is the determination of the nucleotide sequence itself. The original target is that this should be achieved by the year 2005. In this section the strategies and technologies that might be employed are reviewed.

8.3.1 Strategies: cDNA or genomic DNA?

Only about 3% of the human genome encodes proteins. Therefore there is a strong argument that sequencing should concentrate initially on this small fraction of the genome that is likely to be the most interesting. The counter- vailing argument is that such an approach will miss sequences important in gene expression and we will lose the opportunity to study chromosome structure and evolution.

Expressed sequences can be isolated from cDNA libraries. ESTs are prepared from the 3' UTR (see Chapter 3) and used to identify clones in the genomic or cDNA libraries, from which the rest of the coding sequence of a gene can be derived. Alternatively, a large number of short sequences can be derived by PCR from cDNA libraries and then assembled into complete sequences by piecing together the overlaps in the fragments. A large data- base of expressed sequences has been compiled in the USA by a laboratory called TIGR (The Institute for Genomic Research). TIGR is funded by Human Genome Sciences, a commercial company. Other scientists were only given access to this database under conditions which many thought were too restrictive. As a result of the disquiet that resulted, the work is being repeated by the University of Washington, St Louis with support from Merck Pharmaceutical Company. Data from expressed sequences is now deposited in the public domain in the dbEST database (see below).

A major concern with deriving sequences from cDNA libraries is how to be sure that every expressed gene has been identified. cDNA libraries reflect the abundance and pattern of gene expression in the cell from which they were prepared. This raises two problems. The first is that some genes are weakly expressed, i.e. their transcripts represent a small proportion of the total transcript populations. The cDNA clones to which they give rise will therefore be correspondingly under-represented in the library. Secondly, some genes may only be expressed in some cell types and possibly only at restricted times in development. It will be difficult to be certain that every gene has been identified. Complete sequencing of the genome is the only way we can be sure that this has been achieved.

8.3.2 Sequencing technologies

Manual and automatic sequencing based on the Sanger dideoxy protocol
The Sanger dideoxy DNA sequencing protocol, developed in the mid-1970s, is still the basis of most sequencing methods in use today. The original protocol is described in Figure 8.1. This is now normally referred to as 'manual sequencing'. This method was the workhorse of molecular biology laboratories throughout the 1980s. It is a technically demanding procedure, that requires the use of radioisotopes and autoradiography; also experimenters must read and record the results of the autoradiograph themselves. Generally, it would be used on a small scale to sequence individual genes or to check the results of an experimental procedure such as *in vitro* mutagenesis. For this use it was adequate, if somewhat laborious.

From the inception of the Human Genome Project it was appreciated that manual sequencing would not be sufficient to cope with the demands of sequencing the 3000 Mb of the human genome. There was a need for a procedure that did not require unstable isotopes and the attendant autoradiography, and in which the data were collected automatically not manually. This led to the development of automatic fluorescence-based technologies, which still made use of the basic concept of the Sanger method but dramatically reduced both the labour involved and the scope for human error in data collection (after a long period of software development!). The most commonly used system is the commercial Applied Biosytems Prism 377 DNA sequencer (Figure 8.2), which also forms the basis for fluorescence-based microsatellite genotyping necessary for genome scans (Chapter 3).

DNA chips
Chapter 7 showed that DNA chips could be used to compare the sequence of a test DNA with a reference standard. The concept of DNA chips is based on hybridisation of ASOs to the DNA under test. The oligonucleotides are arranged in massive arrays in a small area using a combination of photolithography and oligonucleotide synthesis. The principle of such chips has been demonstrated in detecting mutations in the mitochondrial genome, *BRCA1* exon 11, β-globin gene and HIV protease gene. DNA chips could be used in genome scans for detection of markers linked to susceptibility loci and could also prove useful in the human genome variability project for the rapid detection of polymorphisms.

The use of DNA chips for comparing DNA sequences is now well established. However, in theory DNA chips could also be used to derive sequence *de novo* (Figure 8.3). The principle is to construct a chip contain-

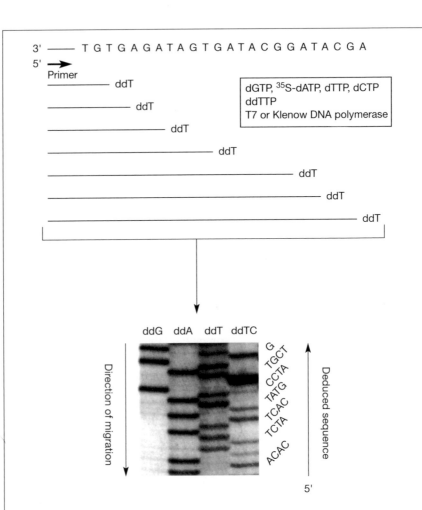

Figure 8.1 Sanger dideoxy sequencing protocol. The single-stranded template is annealed to a primer. The primer is designed to anneal to a particular site in the cloning vector outside of the insert, or may it be designed to anneal to a portion of sequence in the insert that has already been determined. The primer/template partial duplex is incubated with a mixture of the four deoxynucleotides and T7 DNA polymerase, which will extend the primer according to the sequence of the template. Four parallel reactions are carried out in which a different 2',3' dideoxynucleotide analogue is present in each. The dideoxy analogue can be incorporated into a nascent chain but cannot serve as a substrate for the next step in polymerisation. They are therefore said to be chain terminating. In the example shown, dideoxy-TTP (ddT) is included in the reaction mix. At each A residue in the template either dTTP or ddT can be incorporated into the new chain. If dTTP is incorporated, the polymerisation proceeds until the next A in the residue in the template. If ddT is incorporated the polymerisation of that particular chain is terminated. The size of the fragment reports the number of nucleotides from the primer to the ddT causing chain termination. The product is made radioactive by incorporating ^{35}S-dATP in the reaction mix. The products of the four parallel reactions are fractionated by polyacrylamide gel electrophoresis and the molecules visualised by autoradiography. The sequence can be deduced by the order of bands in the four different lanes of the resulting sequencing ladder. On a good day, about 400 bp of sequence can be read from a single gel.

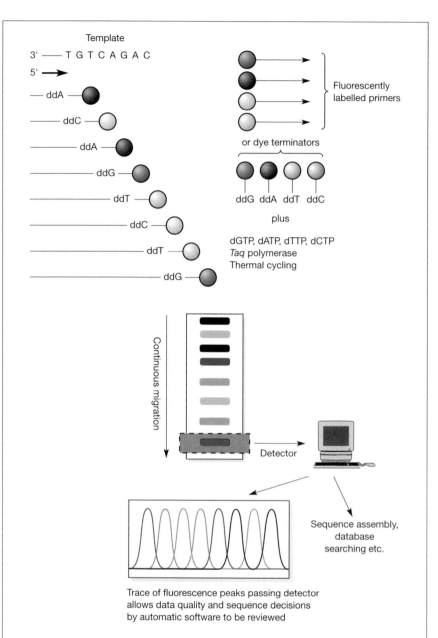

Figure 8.2 Sequencing using the Applied Biosystems 377 sequencer. The sequencing reactions are carried out using either fluorescently labelled primers or fluorescently labelled dideoxy chain-terminating analogues. A thermostable DNA polymerase is used that allows thermal cycling similar to that used in PCR. As a result, the products are amplified in a linear fashion that generates sufficient product to be detected in real time as they pass a detector. If fluorescently labelled primers are used, four independent reactions are carried out in parallel and then mixed and run in the same lane. The products of each reaction can be recognised because the primers are labelled a different colour in each of the reactions. If labelled terminators are used, only a single reaction is necessary containing all four terminators labelled with different coloured dyes.

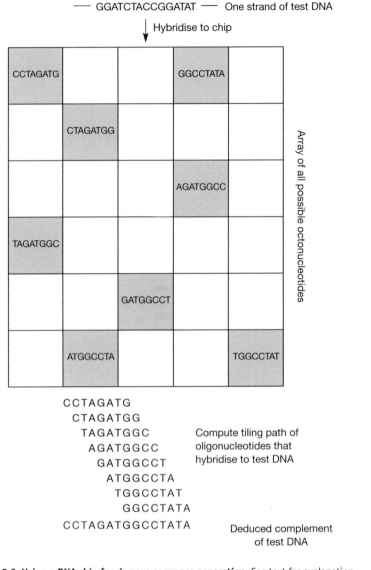

—— GGATCTACCGGATAT —— One strand of test DNA

↓ Hybridise to chip

Array of all possible octonucleotides

CCTAGATG
CTAGATGG
TAGATGGC
AGATGGCC
GATGGCCT
ATGGCCTA
TGGCCTAT
GGCCTATA
CCTAGATGGCCTATA

Compute tiling path of oligonucleotides that hybridise to test DNA

Deduced complement of test DNA

Figure 8.3 Using a DNA chip for *de novo* sequence generation. See text for explanation.

ing every possible DNA octamer (65 536 combinations). The test DNA molecule would hybridise to those octamers on the chip whose sequence matched that of the test DNA. The sequence can then be reconstructed from the **tiling path** of the octamers. Theoretical calculations show that up to 200 bp of sequence could be determined with an array of 65 536 oligonucleotides. At present the problem with this idea is that the duplexes formed between the octamers and test DNA are only stable under non-stringent hybridisation conditions, which also permit intramolecular base pairing in the target DNA. This intramolecular base pairing prevents hybridisation to the oligonucleotides. The only way around this impasse is to use longer oligonucleotides; however, the number of different oligonucleotides that would be necessary to complete the array exceeds the current capacity of existing methods of chip construction. Another problem with DNA chips is that it will be very difficult to cope with repeated sequences.

8.3.3 Preparing sequence-ready clones

The physical maps generated by the first stage of the Human Genome Project were based on YAC clones that have inserts in the megabase size range. These are too large to serve as a template for sequencing. The genome must be fragmented into pieces of more manageable size before sequencing can begin. A high proportion of YAC clones are chimeric and also show structural instability, leading to deletions and rearrangements within the insert. This raises concern about the wisdom of starting the sequencing with such clones. It will be necessary to generate new clone libraries in cosmid, BAC or PAC vectors. Like YAC clones, cosmid clones are also unstable and some human sequences are difficult to clone in this vector. BAC and PAC vectors can carry 100–300-kb inserts and are thought to preserve the topology of the chromosomal sequence more faithfully than YAC vectors.

Whichever vector is used, the STS content map can be used to order the clones. The integrated STS content map, described in Chapter 3, provides a framework map of each chromosome. However, higher-resolution physical maps of individual chromosomes will need to be constructed before sequence-ready clones can be generated (Figure 8.4).

Once the new library is constructed and ordered, the inserts in the cosmid/BAC or PAC vectors will then be fragmented by shotgun cloning into vectors such as M13, which provide single-stranded DNA templates for sequencing (Figure 8.5). These smaller fragments will be sequenced on both strands and the sequence of the parental clone pieced together from the overlap between the smaller sequences. In turn, entire chromosomal sequences will be assembled from the sequence in each cosmid or BAC clone.

An alternative route to sequence-ready clones: BAC-end sequencing
The need to construct high-resolution maps of each chromosome is proving to be a significant impediment to starting genome sequencing. An alternative strategy has been proposed, called **BAC-end sequencing**, which claims to bypass the need for high-resolution maps. The principle is illustrated in Figure 8.4. A new genomic library will be constructed in a BAC vector with about 150-kb inserts. The library will contain about 300 000 clones, enough to provide 15-fold coverage of the genome. The clones would be individually stored in microtitre plates. About 500 bp of sequence will be sequenced from each end of each clone. The sequence identifies the clone and overlaps with other clones whose sequence is known. They are known as sequence tag connectors (STCs). There will be about 600 000 STCs scattered at approximately 5-kb intervals throughout the genome. The STCs

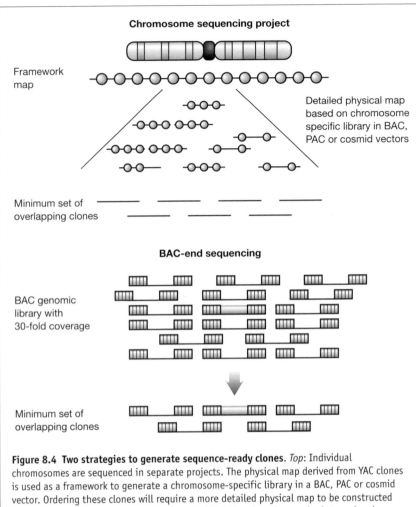

Figure 8.4 Two strategies to generate sequence-ready clones. *Top*: Individual chromosomes are sequenced in separate projects. The physical map derived from YAC clones is used as a framework to generate a chromosome-specific library in a BAC, PAC or cosmid vector. Ordering these clones will require a more detailed physical map to be constructed based on the generation of further STS sites (circles) or by other methods to order clones, such as fingerprinting and hybridisation. Once the clones have been ordered, the set of clones that represents the sequence of the entire chromosome with the minimum overlap (minimum tiling path) will be selected for sequencing. *Bottom*: BAC-end sequencing. A whole genome library with 30-fold coverage is prepared in a BAC vector. The ends of each are sequenced and fingerprinted (boxes with vertical lines). A seed clone is selected and completely sequenced (highlighted in blue). The sequence of the seed clone will match the end sequence of about 30 other clones. The two clones with the least overlap, one at each end, are selected to be the next clones to be sequenced. Matching of the fingerprints provides an additional check of the authenticity of the overlap.

will be made widely available electronically on the Web. Each end will also be identified by a restriction enzyme-derived fingerprint (see section 3.6.1).

Any laboratory or sequencing centre can participate in the sequencing project by sequencing a BAC clone using any suitable strategy, e.g. shotgun cloning into M13 or pUC (Figure 8.5). Since the average insert size will be 150 kb and

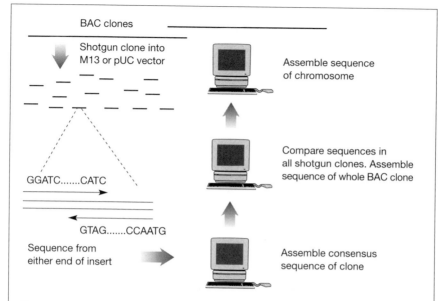

BAC clones

Shotgun clone into
M13 or pUC vector

Assemble sequence
of chromosome

GGATC.......CATC

Compare sequences in
all shotgun clones. Assemble
sequence of whole BAC clone

GTAG.......CCAATG

Sequence from
either end of insert

Assemble consensus
sequence of clone

Figure 8.5 Determining the sequence of a BAC clone by shotgun cloning into M13 or pUC vectors. The insert in the BAC clone is digested with frequent-cutting restriction enzymes, such as *Sau*3A or *Taq*I to produce small fragments that are cloned into a convenient vector, such as M13 or the pUC series of plasmids. This procedure, known as shotgun cloning, gives a random collection of overlapping clones. The sequence of 400–500 bp is obtained from each end of the clones. The sequence of the BAC clone is then pieced together from the fragments of overlapping sequence. Ideally, the whole sequence should be derived from both strands of the DNA molecule. In practice this may not always be achieved, but each part of the sequence should be derived from more than one sequencing reaction to provide a check on accuracy. Finally, the sequences from the overlapping BAC clones are assembled to produce the final sequence of the chromosome.

the STCs are 5 kb apart, the sequence of each BAC clone should allow it to be connected to about 30 other clones. The two clones showing the minimum overlap will be selected to continue the sequence bidirectionally from the 'seed clone'. The fingerprints will be used to provide an additional check that the overlap between the clones is authentic. In this way it is estimated that 20 000 BAC clones will be sufficient to sequence the human genome.

There are a number of advantages to this strategy.

- It removes the need to construct high-resolution physical maps before sequencing can begin.

- The sequence of the 600 000 STCs amounts to 10% of the total sequence in the genome and will be of value for this reason as well as their function in connecting overlapping BAC clones. This sequence can be collected by a large sequencing centre in about 2 years.

- STS and EST markers already mapped will be represented in the STCs as well as the sequence of complete BACs. Thus approximate locations of the BAC clones will quickly become evident. Regions of particular biological interest can be selected for priority attention.

- If this strategy proves efficient then it could become a standard method for sequencing the genomes of other organisms, such as microorganisms of biotechnological interest or important parasites such as *Plasmodium falciparum*.

Pilot programmes to demonstrate the practicality of this approach have been funded by the Human Genome Project. It is estimated that the sequencing necessary to generate the STCs can be achieved in 2 years by a large sequencing centre.

This strategy has been criticised as too simplistic for the human genome. Recurrent repeated sequences, such as the *Alu* elements, will make sequence assembly difficult.

8.3.4 Sequencing centres

Generating detailed physical maps of individual chromosomes, assembling contigs of sequence-ready clones and carrying out the sequencing itself is a large-scale operation beyond the scope of a normal research laboratory. This has led to the evolution of sequencing centres that are staffed and equipped for a factory style of operation. Each sequencing centre will concentrate on one or a small number of chromosomes and usually is also active in mapping and sequencing a model organism. As an example the Sanger Centre is described in Box 8.1. A list of sequencing centres worldwide is given below.

Albert Einstein School of Medicine
Applied Biosciences, Division of Perkin-Elmer
Baylor College of Medicine
Eleanor Roosevelt Institute
Genome Therapeutics Corporation
Institute for Genomic Research
Japanese Science and Technology Corporation (JST) Sequencing Project
Lawrence Berkeley National Laboratory
Lawrence Livermore National Laboratory
Los Alamos National Laboratory
Max-Planck-Institut für Molekulare Genetik
Sanger Centre
Stanford University
University of Oklahoma
University of Texas Southwestern Medical Center
University of Washington
Washington University Center for Genetics in Medicine
Washington University School of Medicine Genome Sequencing Center
Whitehead Institute/MIT Center for Genome Research

Further information may be obtained from **hhtp://www.hugo.gdb.org/ hsmindex.html**

BOX 8.1: THE SANGER CENTRE, UK (http://www.sanger.ac.uk)

The Sanger Centre was established in 1992 as a joint venture between the British Medical Research Council and the Wellcome Trust for Biomedical Research. It is situated near the university town of Cambridge. It has a staff of approximately 310 and so far has placed 50 Mb of DNA sequence in the public domain. In 1995 the Wellcome Trust announced funding of £50 million to enable it to sequence 500 Mb (one-sixth) of the human genome. It will concentrate on chromosomes 1, 6, 20, 22 and X (the sequencing of chromosomes 22 and X are collaborative with the Genome Center, St Louis). The Sanger Centre has also sequenced the following areas of special interest:

Chromosomal region	Size	Reason of interest
3p21.3	0.3 Mb	LUCA6 region
4p	1.6 Mb	HD region
11p13	0.2 Mb	PAX6 region
11p15.5	80 kb	–
12	–	MODY3 region
13q12	0.9 Mb	BRCA2 region
16p	0.3 Mb	β-globin region

As well as its activity in the human genome, the Sanger Centre made a large contribution to the now completed sequencing of *S. cerevisiae* and is currently involved in sequencing the following model organisms:

Bacterial genomes
> *Mycobacterium tuberculosis*
> *Mycobacterium leprae*
> *Neisseria meningitidis*
> *Streptomyces coelicolor*

Plasmodium falciparum (a protozoan organism that is the causative agent of malaria)
Schizosaccharomyces pombe (fission yeast)
Caenorhabditis elegans (roundworm), currently a major project
Drosophila melanogaster (fruit fly)

8.3.5 Informatics

As discussed above, electronic forms of communication and analysis are a key part of the Human Genome Project. Genome maps are maintained in an electronic format that can be accessed and interrogated over the World Wide Web. Addresses of Web sites are listed in Chapter 3.

Databases of protein and amino acid sequences, such as Genbank, EMBL, Swissprot, etc., were established in the early 1980s. They became the first port of call once a protein sequence had been derived for a gene of interest. Very often connections were made between proteins in different organisms that were involved in apparently unrelated biological processes. An early example of this was the demonstration that the protein encoded by the v-*sis* oncogene was a form of platelet-derived growth factor. There have been numerous further examples. A recent one was the demonstration that the human *NF1* gene, mutations of which cause neurofibromatosis, is homologous to the yeast *IRA2* gene, first recognised as causing a sporulation defect (see below). The big pay-off from these connections is that research in one field helps the other. In the case of *NF1* and *IRA2*, it was known that *IRA2* is a negative regulator of a Ras-like protein found in yeast.

As well as similarity at the level of whole proteins, particular amino acid motifs had been defined for particular protein functions, for example NBFs, protein kinase domains, glycosylation sites, DNA-binding domains, etc. Even if a particular protein was not homologous to another, very often clues to its function could be deduced from the presence of these motifs.

As the sequence of the 65 000–80 000 proteins emerges from the Human Genome Project this type of analysis will be essential to understand the function of the proteins. A selection of some of the databases and analytical tools that have been developed is shown in Table 8.1. Most of the Web addresses shown have links to other sites.

8.3.6 *S. cerevisiae*: the first eukaryote genome to be sequenced

The genome of the yeast *S. cerevisiae* was the first eukaryote genome to be sequenced. Apart from a few compact bacterial genomes, it is the first genome of any organism to be completely characterised. This section reviews the *S. cerevisiae* genome project not only for its own intrinsic interest but also as an example of the benefits that flow from sequencing model organisms. There are two basic reasons why the yeast *S. cerevisiae* has proved so useful as a model organism, both of which illustrate the general reasons why model organisms form an essential part of the Human Genome Project. Firstly, fundamental cellular processes are similar in yeast and humans, but in yeast they are less complex and thus easier to understand. Secondly there is an impressive tool kit that can be used to analyse the function of genes in yeast. Some of the more important technical advantages are listed below.

- Yeast is easy and quick to grow. Large numbers of colonies can be conveniently screened in mutant hunts.

- Yeast has a life cycle well suited to conventional genetic analysis. This has allowed the construction of a detailed genetic map, as well as the normal insights gained from the genetic analysis of gene function.

- Genes can be readily deleted from the genome to discover the phenotype of the null mutation. Where gene function is essential, the gene can be deleted in a diploid cell and its indispensability demonstrated by the non-viability of haploid segregants. The expression of essential genes from

Table 8.1 List of electronic resources used to study protein structure and function (a list of mapping sites is presented in Chapter 3).

Resource	Role	URL
Genbank	DNA and protein sequence databases	http://www. ncbi.nlm.nih.gov
EMBL	DNA sequence	http://www.ebi.ac.uk
Swiss Prot	Protein sequence database	http://www.expasy.ch/sprot/sprot-top.html
PIR	Protein database	http://www.gdb.org/Dan/proteins/pir.html
dbEST	Database of sequences derived from human ESTs	http://www.ncbi.nlm.nih.gov/dbEST/index.html
Unigene	Database of single representative EST from each unique gene	http://www.ncbi.nlm.nih.gov/UniGene/index.html
XREFdb	Database of homologues of human disease genes in model organisms	http://www.ncbi.nlm.nih.gov/XREFdb/
SGD	*S.cerevisiae* genome database. Contains links to other yeast genome resources	http://www.genome-www.stanford.edu/Saccharomyces/
Pedros Biomolecular Research Tools	List, with hot-links, of resources to analyse protein structure and function	http://www.public.iastate.edu/~pedro/rt_1.html

regulatable promoters can be engineered. This allows the gene to be turned on and off at will in order to investigate its cellular function. It also allows the effects of overexpression to be investigated. Hypotheses regarding the function of the encoded protein can be investigated by transplacement of the wild-type gene with a gene carrying specific mutations.

- Interactions between proteins can be detected by the so-called two-hybrid screen, which couples protein–protein interactions to a colour assay based on the activation of a β-galactosidase reporter gene. This type of analysis has proved vital in a large number of different investigations into cell function. One particularly important aspect of this assay is that cDNA libraries from other organisms, including humans, can be screened in yeast for interactions between pairs of proteins. It is thus an important research tool for all types of research in cell biology.

- Sometimes pairs of genes are found where the deletion of either can be tolerated, but deletion of both is lethal. Such a situation is called a synthetic lethal interaction. Increasingly, we are coming to appreciate that even in a simple organism such as yeast, cellular function is controlled by networks rather than simple linear pathways. Like the design of the Internet or World Wide Web, damage to part of the system is accommodated by alternative routes; however, damage to multiple parts of the system eventually causes it to fail. Special screens have been developed in yeast to allow synthetic lethal interactions to be detected.

S. cerevisiae genome

The *S. cerevisiae* genome is 12 068 kb in size divided between 16 chromosomes. There are 5885 ORFs, which could encode polypeptides larger than 99 amino acid residues. It is unusually compact for a eukaryote genome, as there is a protein-coding gene every 2 kb. By contrast, the density of genes in the human genome is approximately one every 45 kb. Unusually for a eukaryote, only 4% of *S. cerevisiae* genes contain an intron. Those that do usually have a single small intron near the start of the coding sequence. It has even been suggested that most *S. cerevisiae* genes are retroposons, generated by the action of reverse transcriptase encoded by a transposable element called Ty1 found in the yeast genome.

The simplicity of the yeast genome makes it very easy to identify every potential coding sequence and design experiments to investigate their function. This is in striking contrast to the human genome. As discussed above, identifying every coding region in the human genome will be very difficult.

Yeast proteome

We now know the identity of every protein that the yeast cell is capable of synthesising. This is known as the **proteome**. About 50% of the proteins in the yeast proteome can be ascribed a general function, such as 'protein kinase' or 'transcription factor' on the basis of its amino acid sequence. A rough classification of the likely cellular function of these proteins as a proportion of the total proteome is shown below.

Metabolism (11%)
Transcription (7%)
Protein trafficking and targeting (7%)
Translation (6%)
Membrane transporters (4%)
Structural (4%)
Energy production and storage (3%)
DNA replication and repair and recombination (3%)

The new challenge is to determine the exact function of these proteins, as well as the remaining proteins whose amino acid sequence gives no clue of their function. Conceivably, it may be possible in the future to completely describe the structure and function of a yeast cell in terms of its protein components.

Table 8.2 Yeast genes homologous to human disease genes.

Disease	Human gene	Yeast gene	Gene function in yeast/phenotype
Hereditary non-polyposis colorectal cancer	MSH2	MSH2	Component of mismatch repair binding factor/increased mutation frequency
Hereditary non-polyposis colorectal cancer	MLH1	MLH1	Homologue of *E. coli mut* L /increased mutation frequency
Cystic fibrosis	CFTR	YCF1	ABC transporter/cadmium sensitive
Wilson's disease	WND	CCC2	Copper-transporting P-type ATPase/iron uptake deficiency
Glycerol kinase deficiency	GK	GUT1	Glycerol kinase/glycerol utilisation defective
Bloom's syndrome	BLM	SGS1	Helicase/suppressor of *top3*
Adrenoleucodystrophy, X-linked	ALD	PAL1	Peroxisomal ABC transporter/ required for growth on oleate
Ataxia telangiectasia	ATM	TEL1	Phosphoinositol 3-kinase/short telomeres
Myotubular myopathy, X-linked	MTM1	SCYJR110W	Not characterised
Amyotrophic lateral sclerosis	SOD1	SOD1	Superoxide dismutase/oxygen sensitive
Myotonic dystrophy	DM	YPK1	Similar to protein kinase C / synthetic lethal with *YPK2*
Fanconi's syndrome	CLCN5	GEF1	Voltage-gated chloride channel/ respiratory growth defective
Werner's syndrome	WRN	SGS1	Helicase/*top3* suppressor
Lowe's syndrome	OCRL	YIL002C	Putative inositol polyphosphate -5-phosphatase
Choroideremia	CHM	GDI1	*SEC5* GDP dissociation inhibitor/secretion defective
Neurofibromatosis type 1	NF1	IRA2	Ras-GTPase activating protein/sporulation defective

Yeast homologues of human disease genes

The relevance of this to human research is shown by the statistic that 21% (15/70) of positionally cloned human disease genes have counterparts in the yeast genome (Table 8.2). It can sometimes be difficult to decide whether similarity of amino acid sequence between human and yeast genes really means that they have descended from a common ancestral gene and carry out the same function. An indication that the similarities really are significant is that the phenotype of a yeast *ira2* mutant can be rescued by the human *NF1* homologue. (In yeast genetic nomenclature a recessive allele, usually the mutant, is written in lower case; the dominant, usually wild-type allele, is written in upper case.) All 15 disease genes have a similarity that is statistically more significant than the similarity between *IRA2* and the human *NF1* homologue. Table 8.2 also lists the gene function and mutant phenotypes of the yeast gene; the variety of biological functions involved is striking. It is perhaps a good illustration of the difficulty in predicting the long-term application of pure research. For example, who could have predicted that the analysis of yeast sporulation would eventually be relevant to human neurofibromatosis type 1?

8.4 Summary

- The sequence of the human genome will allow the following:

  - Describe every single protein (and RNA component) responsible for the structure and function of the human body and specify where and when each protein is used. Understanding the role of each protein will occupy biologists for a long time to come. Sequencing model genomes will greatly aid this investigation since they are simpler and more amenable to experimentation.
  - Illuminate the structure of chromosomes and the evolution of the genome.
  - Allow a more comprehensive understanding of the relationship between susceptibility to disease and genotype.
  - Provide a greater understanding of human diversity and the recent history of evolution of human populations.

- The Human Genome Project was initiated by the DOE and NIH in the USA in the late 1980s. The DOE was charged with monitoring the genetic damage caused by radiation and environmental mutagens. The NIH funded biomedical research that was facing up to the difficulties of cloning human genes. Subsequently, research in other countries contributed to the project. HUGO was established to coordinate work internationally.

- A 15-year strategic plan, to run from 1990, was agreed. The goals of the period up to 1998 are:

- ○ A genetic map of resolution 2–5 cM (now achieved).
- ○ A physical map with a resolution of 1 STS/100 kb (nearly complete).
- ○ An expression map (nearly complete).
- ○ Development of sequencing strategies and technologies.
- ○ Study ethical, legal and social issues.
- ○ Develop technology, communication between researchers and public education.

- Sequencing expressed sequences derived from cDNA libraries provides the sequence of the majority of proteins, but the entire genome will have to be sequenced to be sure that all protein-coding sequences have been identified.

- The original Sanger dideoxy sequencing protocol is too inefficient for large-scale sequencing. It has been superseded by automatic sequencing technology based on primers or dideoxy nucleotides labelled with fluorescent dyes. New technologies based on DNA chips could make sequencing still more efficient.

- Dividing up the genome into sequence-ready clones is a major undertaking. The conventional strategy is to use high-resolution physical maps to order clones in vectors such as BACs, PACs or cosmids. An alternative strategy, called BAC-end sequencing, may bypass the need for high-resolution physical maps.

- The inserts in the BAC, PAC or cosmid vectors will be sequenced after shotgun cloning into vectors such as M13 or pUC.

- The sequencing itself will be carried out in a small number of factory-style centres.

- The first eukaryote genome to be sequenced was that of the yeast *S. cerevisiae*. Its cellular organisation and molecular processes are remarkably similar to those of higher organisms and many genes are conserved; 15/70 human genes that have been positionally cloned have yeast counterparts. These genes are involved in a wide variety of cellular processes.

Further reading

General
CASKEY, C.T. AND ROSSITER, B.J.F. (1992) The human genome project: purpose and potential. *Journal of Pharmacy and Pharmacology*, **44** (Suppl. 1), 198–204.

HUGO Web site
http://www.ornl.gov/TechResources/Human_Genome/research.html

A large resource with an online journal, *Human Genome News*, containing many articles and news of the project.

History and goals
COLLINS, F.S. AND GALAS, D. (1993) A new five year game plan for the U.S. genome project. *Science*, **262**, 43–46.

CANTOR, C.R. (1992) Orchestrating the human genome project. *Science*, **248**, 49–51.

Yeast genome project
BASSET, D.E., BOGUSKI, M.S. AND HIETER, P. (1996) Yeast genes and human disease. *Nature*, **379**, 589–590.

GOFFEAU, A., BARRELL, B,G., BUSSEY, H. et al. (1996) Life with 6000 genes. *Science*, **274**, 546–567.

Sequencing strategies
HEYNINGEN, V. (1996) Changing tack on the map. *Nature Genetics*, **13**, 134–137.

VENTNOR, J.C., SMITH, H.O. and HOOD, L. (1996) A new strategy for genome sequencing. *Nature*, **381**, 364–366.

Describes BAC-end sequencing.

DNA chips for sequencing
Southern, E.D. (1996) DNA chips: analysing sequence by hybridisation on a large scale. *Trends in Genetics*, **12**, 110–115.

CHAPTER NINE

Human population genetics and evolution

Key topics

9.1 Introduction

Apart from monozygotic twins, the genetic constitution of each individual is different. What sets us apart from each other is not that we have different genes – we all have copies of the same genes – but rather these genes may take different forms (**alleles**). In some cases, different alleles may produce different phenotypes, as in mutations responsible for monogenic disorders; in others, the variations may not have any discernible effect on the phenotype and thus are said to be **neutral alleles**.

If a gene or locus exists in more than one allelic form, it is said to be polymorphic. We can measure the frequencies of the different alleles of polymorphic genes and thus determine population **allele frequencies**. Having derived an allele frequency in one population, we can ask whether the frequency is similar in another population. In many cases it is found that they can be very different. We have already seen examples of this, where the frequency of genetic diseases can vary in different parts of the world. We can also ask whether a population allele frequency is stable or changing through time. If it is changing we can investigate the causes.

These questions form the basis of human population genetics which we consider in this chapter. We shall see that the quantitative analysis of gene frequencies allows the reconstruction of the spread of anatomically modern humans from their origin in Africa 100–200 kya (thousand years ago), and reveals the genetic consequences of subsequent events such as the development of agriculture and the intercontinental migrations of the last 500 years. These studies tell us much about the present-day distribution of genetic disease and the normal variation that might contribute to complex diseases.

9.2 Factors that influence population gene frequencies

9.2.1 Hardy–Weinberg distribution.

Consider a gene that has two alleles, say A and a, with the A allele being dominant over the a allele. These alleles can combine in different ways to produce three different genotypes: AA, Aa and aa. In a population at equilibrium, i.e. where the gene frequencies are not changing, the distribution of these genotypes is described by the Hardy–Weinberg equilibrium which states that:

$$p^2 + 2pq + q^2 = 1$$

where p is the frequency of the A allele, q is the frequency of the a allele and $2pq$ the frequency of Aa heterozygotes. This distribution is only valid if a number of conditions are met:

1. mating is random;
2. there is no selection for or against any of the genotypes;
3. there is no migration.

In human populations these conditions are rarely completely satisfied; in particular, mating is rarely random. Humans tend to mate with partners from similar religious, cultural and socioeconomic backgrounds. We shall consider the effects of migration and selection below. Nevertheless, in the absence of obvious factors, such as clear selection for one of the genotypes or large-scale migration, allele frequencies do generally conform to that described by the Hardy–Weinberg distribution.

The usefulness of the Hardy–Weinberg distribution is that if we measure the frequency of one of the genotypes, it allows the other two to be calculated. For example, the average frequency of CF in European populations is about 1 in 2000. Since CF is an autosomal recessive condition, 1/2000 is the value of q^2 in the Hardy–Weinberg distribution. This allows the following simple calculation:

$$q^2 = 1/2000$$
$$q = 1/44$$
$$2pq = 1/22 \text{ (approximating } p \text{ to 1)}$$

$2pq$ is the frequency of heterozygotes, so starting with the observed frequency of the disease, we can calculate the frequency of carriers. We can perform a similar calculation with autosomal dominant traits, but this time the observed incidence of the disease is $2pq + q^2$.

9.2.2 Changes in allele frequency

Allele frequencies do not always remain constant. If they did, then evolution could not have occurred, since evolution requires such changes. New alleles can only be created by mutation. The classic neo-Darwinian theory of evolution states that alleles change in frequency in response to **selection**, either for advantageous alleles or against deleterious alleles. However, selection is not the only mechanism that operates; migration and random drift of selectively neutral alleles also act to change allele frequencies.

Selection
Selection is a consequence of alleles that alter the **Darwinian fitness** of the organism, defined as the number of its offspring reaching reproductive maturity. The rate of change in the frequency of an allele in response to selection can be calculated from its selection coefficient, which is the fractional difference between Darwinian fitness of an organism carrying the allele and the Darwinian fitness of a reference genotype. Selection is a powerful force when it operates against strongly deleterious alleles in a population, such as those responsible for monogenic disorders. If such a disorder has a selection coefficient of 1, i.e. no affected individuals have children, then selection will ensure that the frequency of homozygous individuals affected by an autosomal recessive disorder will be equal to the frequency with which new alleles are generated by mutation. There is an important corollary to this conclusion: if the frequency of a genetic disease in a population is higher than would be consistent with a reasonable rate of mutation, then some factor must be operating to maintain the allele frequency. We consider several such situations in this chapter.

The essence of the theory of Darwinian evolution is that favourable mutations will increase in frequency due to selection. The rate at which a beneficial allele would spread through a population can be calculated from the selection coefficient. One of the best-known examples of positive selection is the **heterozygous advantage**, in areas where malaria is endemic, conferred by heterozygosity for sickle cell anaemia and other mutations that affect haemoglobin (discussed in more detail below). The experimentally measured selection coefficient for carriers of such alleles is about 0.1.

Allele frequencies of genes involved in resistance to infectious diseases, such as those of the immune system, are highly variable in different parts of the world, which is assumed to reflect different histories of exposure to infectious diseases. As discussed below, modern humans probably emerged from Africa between 50 and 100 kya. If selection had been operating throughout this time, the value of selection coefficients operating on genes of the immune system would be in the order of 0.01.

Random drift

Mathematical analysis, by geneticists such as Fisher and Haldane, showed that the effects of selection on Mendelian genes provides a satisfactory mechanism for evolution. This became known as the neo-Darwinian synthesis, and for a while it was thought to be the only mechanism responsible for changes in allele frequency. An important implicit assumption of the neo-Darwinian synthesis is that all, or nearly all, genetic variation is subject to selection: an allele is either advantageous or disadvantageous but rarely neutral. During the 1960s, the molecular analysis of gene structure and evolution started to reveal a number of facts that were not readily explicable in terms of selection. Two of the most important observations were:

1. At the level of the amino acid or nucleotide sequence most genes were much more variable than predicted by the neo-Darwinian theory. This fact emerged during the study of protein polymorphisms revealed by electrophoresis. At first it was hotly contested, but the advent of DNA sequencing has showed that this statement is undeniable.
2. The rate of nucleotide or amino acid substitution was apparently constant in a wide variety of genes examined. This gave rise to the concept of the molecular clock (Figure 9.1).

The first observation was inconsistent with the neo-Darwinian view, because if few alleles were selectively neutral, selection should ensure that the most advantageous allele became fixed in the population. Most genes should therefore be identical. Polymorphisms should be rare and only observed when the frequency of a new and advantageous allele was being increased by selection. The second observation is surprising because it would be expected that proteins would evolve to a configuration that optimised their function and that subsequent changes would be selected against. It is difficult to imagine how, over long periods of evolution, new alleles could arise and be selected for with the same selection coefficient.

These observations led Motoo Kimura to propose what is now known as the **neutral theory of evolution**. The foundation of this theory is that many, or even most, alleles are selectively neutral. Allele frequencies change

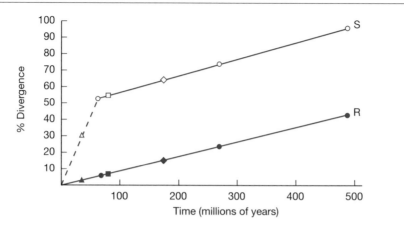

Figure 9.1 Molecular clock of globin evolution. The sequences of α-globin and β-globin genes were used to measure the divergence between pairs of globin genes. Both silent substitutions (filled symbols) and replacement substitutions (open symbols) accumulate at a constant rate. At first, the rate of silent substitutions was 10 times that of replacement substitutions, in agreement with the neutral theory of evolution. Eventually all possible selectively neutral substitutions that could occur had occurred. Further substitutions were subject to the same selective pressures as replacement substitutions and the rate at which substitutions accumulated slowed down. The data were obtained from the following comparisons, which can be dated from the archaeological record (circles): intraspecies comparisons of α- and β-globin genes which were presumed to have diverged 500 mya (million years ago); comparisons of α- and β-globin genes of mammals with those of chicken (270mya), α- and β-globin genes of different mammals (85 mya). Further data for which archaeological dates are not available were obtained from comparison of δ versus β (triangles), γ versus ε (squares) and δ or β versus γ or ε (diamonds). For these data the replacement substitutions were plotted on line R and the silent substitutions plotted directly above. The amount of divergence (corrected for multiple substitutions) was plotted against the time since divergence. Line R passes through the origin. Because line S does not pass through the origin and there is no independent verification of the time of the β/δ divergence, it is plotted as a dashed line.

because of random sampling effects between generations. This can be explained by a simple analogy. Consider a bag containing equal numbers of white and red balls. If 10 balls were randomly withdrawn from the bag, it would not be surprising if the sample consisted of four red and six white balls. The frequency of red balls has thus changed from 0.5 to 0.4. However, if we withdrew 1000 balls we would be very surprised to find that the sample contained 400 red and 600 white balls. If the different coloured balls represent two alleles of a gene, two conclusions are evident from this analogy. Firstly, random sampling will result in a change in allele frequency from generation to generation. Secondly, the smaller the population size, the larger the changes are likely to be.

At every generation the frequency of an allele will fluctuate due to this sampling effect until either the frequency of an allele becomes 100% (fixation) or it disappears completely from the population (extinction). One of these two outcomes is logically inevitable; the time it takes depends on population size, occurring more rapidly in smaller populations. Clearly, fixation is a more likely outcome for common alleles and extinction more likely for rarer alleles. Thus the most likely fate for a new allele introduced by mutation is that it will become extinct. However, since a large number of alleles are continually being created by mutation, fixation of some of them will occur by random drift.

It is important to understand that the neutral theory does not state that all alleles are neutral. Firstly, rare advantageous alleles do exist and will increase in frequency due to selection. However, they are likely to be rare and make little contribution to overall variability. More importantly, many substitutions will be deleterious. Kimura showed that the rate of evolution can be described by the following equation:

$$k = v_T f_0$$

where k is the rate of nucleotide substitutions per year, f_0 the fraction of new alleles that are selectively neutral and v_T the mutation rate. This equation infers that the rate of evolution of a gene, or part of a gene, will be constant and will depend on two variables, the mutation rate and the fraction of selectively neutral mutations. The validity of this equation is readily tested by comparing the rate of nucleotide substitutions in situations where we may be confident about whether or not the result is likely to be selectively neutral. For example, Figure 9.1 compares the rate of two different types of nucleotide substitutions that have occurred in the evolution of the globin genes. **Silent substitutions** do not result in a change in amino acid sequence, because they result in a synonymous codon, i.e. a codon that encodes the same amino acid. **Replacement substitutions** result in a codon that encodes a different amino acid. Clearly, silent substitutions are more likely to be selectively neutral than replacement substitutions. Figure 9.1 shows that this is indeed the case. The initial rate of silent substitutions occurs at a rate of 1% divergence per million years, replacement substitutions at about one-tenth of that rate. Silent substitutions in coding regions may still be subject to some selective pressure, for example the secondary structure of the mRNA may be altered. Sequences that are unlikely to be subject to any selection, such as pseudogenes, evolve at a rate of 2% divergence per million years.

Genetic drift and selection are therefore factors that act together: random drift ensures that substitutions occur at a constant rate, while selection controls that rate. Neutral alleles evolve at a rate determined by the mutation frequency; protein sequences evolve at a lower rate, determined by the capacity of the protein to accommodate amino acid substitutions. Different proteins evolve at different rates as a result of this constraint. Even different parts of proteins may evolve differentially, for example the active site of an enzyme is likely to be more highly conserved than other parts of the protein. A last point to note is that selection and drift may operate simultaneously, so that rare advantageous alleles may still become extinct by genetic drift.

The effects of drift and selection described above apply formally to isolated populations. Few populations are entirely isolated: immigration/migration between adjacent populations will result in gene flow that will affect their genetic structure, in particular mitigating the effects of drift. Gene flow between populations as a result of migration is called **admixture**. Migration may also include relocation over a long distance, either to inhabit a previously uninhabited territory or to displace or mix with another previously unconnected population. A new group settling in a previously uninhabited territory will become a **founder population**. Founder populations are likely to be smaller than the parent population from which they were derived and are isolated from them. This will result in the genetic structure in the population derived from the founder group being different from the ancestral population. Firstly, the small size of the founder group means that its gene pool will be less diverse than the parental population. Secondly, by chance, the founder group may have a higher frequency of some rare alleles and, conversely, some rare alleles may be completely absent. Thirdly, since the founder group is small, and will remain small for at least several generations, the effect of genetic drift will be large.

It is important to realise that religious, cultural and linguistic barriers may inhibit gene exchange between populations that live in the same geographic location, so the founder effect commonly affects ethnic and cultural communities after they have migrated to a new country. The Ashkenazi Jews and Amish communities in the USA are clear examples of this. In both groups the frequencies of certain genetic diseases are much higher than observed in other populations (see below).

Population bottlenecks will have similar consequences to the founder effect. A bottleneck is caused when a previously large population contracts to a small number and then expands to its previous numbers. As a result the gene pool may be less diverse and gene frequencies may be altered because of the effects of sampling and genetic drift. Bottlenecks in human evolution may have occurred as a result of famine or epidemics of infectious disease.

9.3 Human genetic variability

Definition of a population

We have already seen that individual humans are genetically variable. This variability extends to a higher level, i.e. gene frequencies differ in different human populations. Before we explore this statement in more detail, we should consider what is meant by a population. Genetically, a population is a group of individuals who mate randomly with each other. In such a population the frequency of genotypes at polymorphic loci will conform to the Hardy–Weinberg distribution. Barriers to mating will partition the world's population into subpopulations. The most obvious of these barriers is geographic distance, together with features such as mountains and seas. Even groups that live in geographic proximity do not always interbreed.

Additional barriers may be language and cultural affiliation (tribe, country, etc.). In practice none of these barriers is absolute and it is rarely possible to define a population unit that is sharply distinguished from a geographically adjacent one. For the most part, genetic features that distinguish one population gradually merge into the next population. Sometimes a population will appear to be sharply defined by one allele frequency, whereas another allele frequency may yield an entirely different picture. Despite these problems we can define populations on the basis of geographic, cultural and linguistic criteria. Measurements made in the centre of such populations are likely to represent real differences between them, although at the periphery they are likely to merge into one another.

Phylogenetic trees

To be useful the large amount of data that results from allele frequency measurements must be synthesised into a form that summarises the differences between populations and is informative about their evolution. This may be done in two ways. The most common method is to construct a **phylogenetic tree** that attempts to describe the evolutionary relationships between the populations. The phylogenetic trees described here are all rooted, i.e. all existing populations are portrayed as having descended from a common ancestral population. The order in which the branches split describes the order in which populations diverged. In some methods the length of the branches diverging from a single node represents the evolutionary distance between populations on each branch.

There are a number of ways that trees may be constructed (a full discussion of these is outside the scope of this book but see Further reading for a description). Basically they may be divided into two categories: numerical or distance methods compare quantitative measures of the similarity or difference between populations; **cladistic** or character state methods use qualitative differences to re-create the actual path of evolution from a common ancestor. There has been fierce and often acrimonious controversy between the adherents of the different methods. However, all agree that none of the methods directly reconstructs the true tree; rather they are a statistical inference of the true tree. For this reason phylogenetic trees should always be regarded with some caution.

The data on allele frequency in different human populations have been mainly analysed by a numerical method called unweighted pair group method with arithmetic mean (UPGMA). The underlying assumption of this method is that genetic distance is a measure of the time since two populations diverged from a common ancestral population. The genetic distance between populations is measured as a function of differences in allele frequencies. The population pair with the smallest distance between them is taken as marking the lowest split in the tree. These two populations are pooled and the process repeated until there are only two populations remaining. This method was chosen for a number of reasons.

- It is computationally straightforward.

- It produces a rooted tree, i.e. it shows all populations as having descended from a common ancestor.

- It leads directly to the tree that best describes the data so there is no need to apply further statistical tests to choose between alternatives.

- Simulations showed that it predicted the correct tree as often as other more complex methods.

An important assumption of this method is that the rate of evolution has been constant. However, because most human populations have undergone extensive size fluctuations in the past this assumption is unlikely to be true. Another criticism is that the method takes no account of population admixture. Indeed the fact that populations merge as well as split means that there may not even be a unique phylogenetic tree.

An alternative method of constructing phylogenetic trees is called maximum parsimony. This method searches for the tree that postulates fewest evolutionary events to explain the observed differences between populations. The method does not produce a unique tree, so it is necessary to carry out a statistical test called a bootstrap analysis in which the analysis is iterated, i.e. on each occasion the dataset is randomly sampled and a small number of data are randomly replaced with others from the dataset. The percentage number of times the computer program generates the same tree is a measure of confidence in the result.

Synthetic maps

The second way of summarising data from different populations is called **principal component (PC) analysis**. It mathematically combines the data from all the experimentally determined gene frequencies and produces a **synthetic map** depicting their geographic variation. PC analysis cannot represent all data simultaneously. It is necessary to represent the data in a series of synthetic maps based on the first, second and third PCs, each succeeding map using the information not used by the preceding maps. The first and second PC maps together represent about 50% of the total information and are a convenient way of visualising multivariate data from hundreds of genes.

9.3.1 Polymorphisms of protein-coding genes

If allele frequencies are measured in different population groups throughout the world, it will often be found that they vary considerably. The first observation of this kind was made in 1919 with ABO blood groups (Figure 9.2).

Subsequently, systematic measurements have been carried out using a wide variety of polymorphisms that can be detected at the protein level, usually by electrophoresis, such as blood groups, serum proteins, HLA, enzymes and other readily detectable proteins. A large and comprehensive study of such markers has been carried out by Luca Cavalli-Sforza and

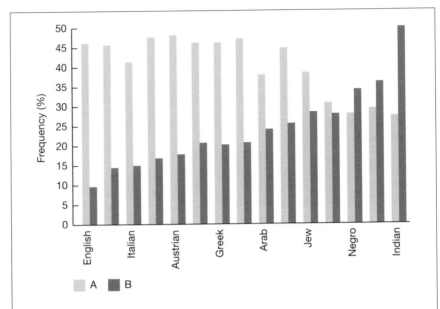

Figure 9.2 Worldwide variation in A and B blood groups. The ratio of people with blood group A to those with blood group B varies from 4.5:1 in the English population to 0.5:1.0 in the Indian population.

colleagues using 120 allele frequencies in 42 populations worldwide. In this context 'populations' refer to populations that are probably still similar to those that existed in 1492, before the mass migrations of the last 500 years. We discuss the consequences of these mass migrations in more detail below. The phylogenetic tree derived by the UPGMA method (see above) is shown in Figure 9.3 and a geographical representation of the worldwide variation in allele frequencies is shown in Figure 9.4.

The phylogenetic tree suggests that the greatest difference in human populations is between Africans and non-Africans. This is a major conclusion that suggests that modern humans originated in Africa and thereafter spread to the rest of the world. This conclusion is reinforced by examination of the synthetic map based on PC analysis (Figure 9.4). On the basis of genetic differences, African populations are clearly differentiated from the rest of the world.

The second branch of the phylogenetic tree shows that non-Africans split into North Eurasian and South East Asian branches probably about 60 kya. The North Eurasian branch gave rise to Caucasians, North East Asian and American peoples. The South East Asian branch gave rise to the populations of New Guinea, Australia and Polynesia as well as South East Asia. The exact order of branching in this South East Asian group does vary according to the method used to measure genetic distance. One tree shows the New Guinean and Australian populations splitting off from the main tree before the split that gave rise to South East Asians.

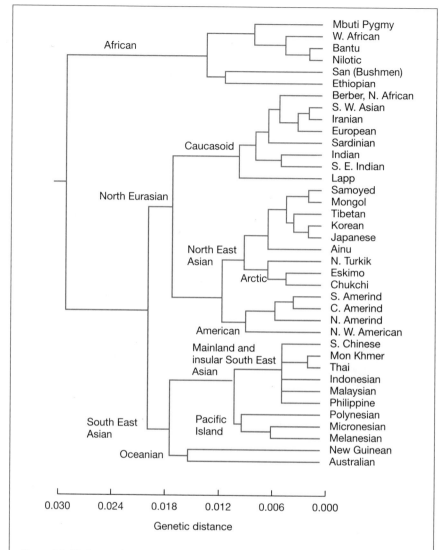

Figure 9.3 Phylogenetic map of human populations based on genetic distances. The order in which the Oceanian and South East Asian divergence occurred is uncertain. When the genetic distances between the populations are calculated in a different way, the Oceanian group splits off before the South East Asian group.

    Three other general conclusions become apparent from examination of this data. Firstly, as a species *Homo sapiens* is genetically uniform compared with a sister species such as chimpanzee. This suggests a relatively recent origin of humankind from a small founder population (probably about 10 000 individuals). Secondly, the variation between populations used as the basis for the phylogenetic tree is nevertheless small compared with differences within populations. Thirdly, the phylogenetic tree based on

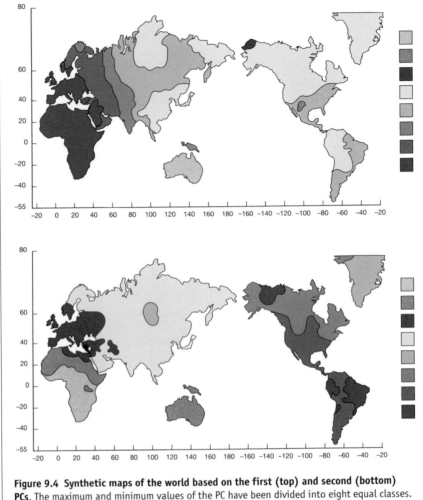

Figure 9.4 Synthetic maps of the world based on the first (top) and second (bottom) PCs. The maximum and minimum values of the PC have been divided into eight equal classes. The greatest difference in the second PC map is between Australasians and Americans. This reflects the second split in the human tree which separated Australasians from the rest of non-Africans. Together the first and second PCs represent about 50% of the total variation.

these nuclear polymorphisms is not correlated with an anthropometric tree based on dimensions of body parts. These characteristics correlate better with climate and suggest that they have evolved recently in response to localised selection. For example, pale-coloured skin could improve vitamin D production in the low levels of sunlight found in northern latitudes.

It follows that attempts to classify humans into different races that have different innate abilities with respect to intelligence or social characteristics have no scientific basis. We are a genetically uniform species; such differences that exist allow us to trace the emergence of human populations from Africa, but they are insufficient to cause systematic differences in these

innate abilities. Even if they were, the characteristics we might intuitively
use to classify races do not correspond to genetic classifications based on a
wide variety of nuclear polymorphisms.

9.3.2 Mitochondrial DNA polymorphisms

Variation in mtDNA has proved extremely useful in studying recent human
evolution. There are a number of reasons for this.

1. The mitochondrial genome is maternally inherited and there is no
    recombination between paternal and maternal genomes. This makes the
    construction of phylogenetic trees straightforward, because in the
    absence of recombination the number of nucleotide differences between
    two mitochondrial genomes is a direct reflection of the genetic distance
    that separates them. In other words, the trees are based on the history
    of the mtDNA molecule itself and not the histories of populations,
    which is the case with trees drawn from nuclear polymorphisms. As
    described above, studies based on populations may be distorted by
    admixture and variation in size.
2. In contrast to nuclear genes, the copy number of mtDNA is high. This
    makes it easy to isolate DNA and analyse polymorphisms.
3. The sequence of the mitochondrial genome was determined in 1981.
    Because the sequence is so well known it is easy to identify polymor-
    phisms.
4. mtDNA evolves five to ten times more rapidly than nuclear DNA.
    This makes it suitable to study recent evolution, because a sufficient
    number of mutations will have arisen to permit the analysis of popula-
    tion relationships.
5. The phylogenetic analysis was carried out by maximum parsimony
    method, which many consider to be a superior method because it is
    cladistic, i.e. the course of evolution is deduced directly from the differ-
    ence in qualitative characters (see above).

Measurements of mtDNA evolution are usually based on the sequence of
the control region (see Figure 1.7), which shows greater variation than the
rest of the molecule. The studies led to a simple conclusion: all extant
mtDNA genomes descend from a single common ancestor who lived in
Africa about 150–300 kya.

The single common ancestor has been popularised as the 'African Eve'.
This appellation has led to the popular misconception that all humans
descend from a single woman. Although all human mitochondrial lineages
trace back to a single individual woman, this does not mean that she was
the only woman alive at the time when she lived. What it does mean is that
all other mitochondrial genotypes that existed in the other women alive at
the same time have become extinct. This is shown graphically in Figure 9.5.
It can be demonstrated that it is a logical necessity that all mitochondrial
genomes descended from a single woman. A full exposition of the argument

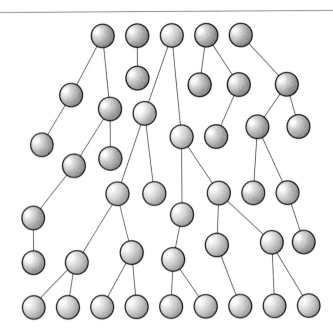

Figure 9.5 The mitochondrial Eve was not the only woman alive. All members of a present-day population ultimately inherit their mitochondria from a single female ancestor as shown by the blue circles. However she was not the only woman alive at the time she lived.

may be found in the references given at the end of this chapter. However, there is no logical reason why the mitochondrial ancestor has contributed any of her nuclear genes to present-day populations.

The origin of the most recent common mitochondrial ancestor in Africa is based on three interrelated lines of evidence. Firstly, phylogenetic analysis produces trees with two branches (Figure 9.6). One branch contains only African populations, the other contains African and non-African populations. The simplest interpretation of this observation is that the ancestor was African. The phylogenetic trees drawn up from these data were controversial for a while, because it was shown that other types of trees would also be consistent with the data. Further work suggests that the original tree shown in Figure 9.6 is probably correct.

The second line of evidence for an African origin is that all mitochondrial lineages found outside Africa are a subset of lineages found within Africa. By far the simplest explanation for this observation is that non-African populations are derived from a group that migrated from Africa. The third line of evidence is that the diversity, measured as the number of different mutations in the African populations, is much higher than non-African populations. Normally the population that has the greater genetic diversity is ancestral to the population with less diversity.

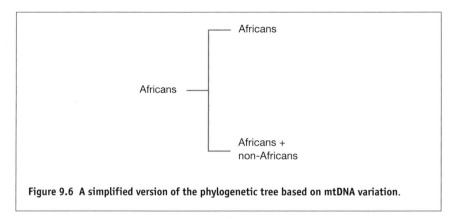

Figure 9.6 A simplified version of the phylogenetic tree based on mtDNA variation.

There are other explanations for these observations. The greater diversity in African populations may reflect a higher rate of evolution or a larger population size, both of which would be expected to increase diversity. The phylogenetic tree may be distorted by the influence of selection. However, the simplest explanation is that modern humans originated in Africa. Moreover, this conclusion is consistent with the large survey of nuclear polymorphisms described above and the analysis of minisatellites described below.

The age of the African ancestor is important because it is critical in distinguishing between the out-of-Africa and the multiregional evolution models of human history (see below). The age of the common ancestor can be simply calculated if two parameters are known: the amount of variation and the rate at which new mutations are generated. The amount of variation was measured experimentally in the original studies and is not controversial. The rate of mutation is more difficult to measure. It is calculated from the amount of mitochondrial variation between humans and chimpanzees using the fossil record to estimate the time of divergence between the humans and chimpanzee lineages. The problem is that the differences between humans and chimpanzees must be corrected for multiple insertions at the same nucleotide site and there is no single reliable method for doing this. There have been a number of different estimates of the age of the mitochondrial ancestor. The estimates range from 100 to 500 kya but most place the divergence between 150 and 300 kya. If true, this figure rules out the regional evolution hypothesis of human evolution, which requires the age of the common ancestor to be over 1 million years old. Recently, direct measurements on the rate of mitochondrial mutation in real pedigrees has found that the actual mutation rate is higher by a factor of at least 20 compared with estimates based on evolutionary comparisons. Thus the rate at which mitochondrial mutations are fixed in the population is lower than that which would be predicted from the rate at which they occur in individual pedigrees. Presumably, drift or selection controls the rate of mitochondrial evolution. This is a further complicating factor in deciding the true rate of evolution.

9.3.3 Minisatellite diversity

Chapter 2 showed that some minisatellites were highly polymorphic for the number of repeat units. This variation formed the basis of the original DNA fingerprinting procedure. The repeat units are not always completely identical; sometimes repeat units in minisatellites can exist in two forms that differ from each other by one or a small number of nucleotide substitutions. The interspersion pattern of these repeat units within the minisatellite locus can also be polymorphic. This form of polymorphism, called intra-locus repeat variation, was originally recognised in the minisatellite locus D1S8, which consists of 29 bp subunits that are repeated to produce a polymorphic locus varying in size from 1.7 kb to 18 kb. Approximately half the subunits differ from the other half by a G to C transition that creates an *Hae* III restriction site. There are thus two types of repeat unit: a-type, which contains the *Hae* III site, and t-type, which does not. The arrangement of these repeat subunits can be determined by a PCR-based procedure called **minisatellite variant repeat PCR** (MVR-PCR; Figure 9.7). This information can then be represented as a digital code.

MS205 is another minisatellite locus that shows intra-locus repeat variation. Analysis of the interspersion pattern of variant repeats in MS205 has provided further evidence for the African origin of modern humans. MS205 is a small locus and all alleles so far examined are less than 5 kb. However it is highly variable: the germline mutation rate has been estimated to be 0.4% per gamete, which is extremely high. However, the process is highly polar, new mutations being confined to five repeat units at one end of the locus. Away from this unstable end, the locus shows reduced variation. This makes MS205 a useful tool to study evolution, because the unstable end can be used to study variation within populations and even in single extended multigenerational pedigrees, while the more stable portion allows variation between populations to be compared and evolution on a longer time-scale to be studied. Figure 9.8 shows examples of the variation found at this locus, illustrating that alleles which differ at their 3' end, can be classified into groups using the more stable pattern found at the 5' end of the satellite. Analysis of populations worldwide has allowed 229 of 242 alleles examined to be assigned to 44 different groups; 10 of these groups contained a particular motif: □■□□■□■□■□. Alleles found in non-African populations were almost exclusively from groups that contained this motif, whereas African populations contained these 10 groups as well as the remaining 34. Again, this study reaches the same conclusion as studies using protein polymorphisms and mtDNA variation: genetic variation in non-African populations is a subset of greater variation found in African populations.

9.3.4 A short history of humankind

The genetic studies outlined above can be combined with the palaeontological and archaeological records to reconstruct the emergence of anatomically modern humans. The fossil record shows that the lineage that

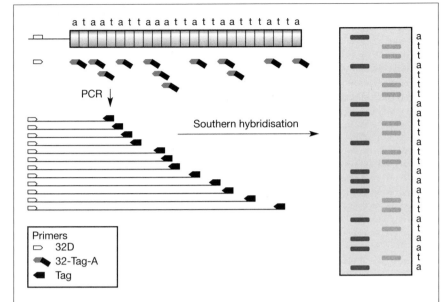

Figure 9.7 MVR-PCR analysis of intra-locus repeat variation. Two types of repeats are found (a-type and t-type) in the minisatellite locus D1S8 (originally recognised by the probe MS32), shown at the top of the diagram as filled and unfilled boxes. The sequence of each type forms a binary code that can be read using MVR-PCR. Each repeat type is examined by two separate PCR reactions that proceed simultaneously. Three PCR primers are used. One of them (32D) anneals outside the minisatellite locus. A second (32-Tag-A) anneals to one type of repeat unit. In the first PCR reaction the sequence between 32-Tag-A and 32D is amplified; 32-tag contains a 5' extension called 'Tag' shown as the black tail. When the molecule produced from the 32-tag-A primer serves as a template for amplification by 32D, the 3' sequence of the product will be complementary to the Tag primer. This allows the product of the first PCR reaction to be amplified using the Tag and 32D primers. Primers 32D and Tag are present at a much higher concentration than the primer 32-Tag-A. The reason for this procedure is that the primer 32-Tag-A could anneal internally to previously amplified PCR products, which would result in the products becoming successively smaller. Using the primer Tag at high concentrations ensures that subsequent amplifications occur mainly from the ends of the molecules amplified in the first round of PCR amplification. Amplification of the t-type repeats is achieved in a similar fashion using 32-Tag-T primers for the initial amplification.

gave rise to modern *Homo sapiens* started with *Homo habilis* in Africa approximately 2.5 mya (million years ago), followed by *Homo erectus* who lived 1.3 mya to 300 kya, who spread from Africa to Europe and Asia. The first members of the species characterised as *Homo sapiens* are found in Africa approximately 400 kya and had spread to Europe by 300 kya. These first human populations are recognisably different from modern humans and are thus called **archaic humans**. In Europe archaic humans formed a subspecies called *H. sapiens neanderthalensis* or Neanderthal man. The last Neanderthals disappeared from Europe 30 kya. Anatomically modern humans are first found in Africa just over 100 kya. Their remains have been found in Israel dated 100 kya, but there is no evidence of modern humans

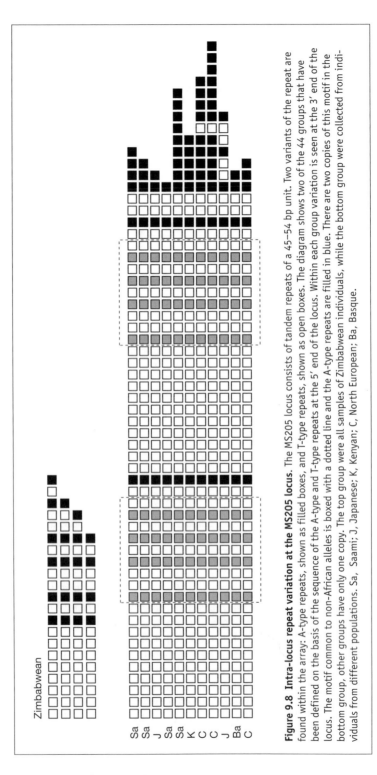

Figure 9.8 Intra-locus repeat variation at the MS205 locus. The MS205 locus consists of tandem repeats of a 45–54 bp unit. Two variants of the repeat are found within the array: A-type repeats, shown as filled boxes, and T-type repeats, shown as open boxes. The diagram shows two of the 44 groups that have been defined on the basis of the sequence of the A-type and T-type repeats at the 5′ end of the locus. Within each group variation is seen at the 3′ end of the locus. The motif common to non-African alleles is boxed with a dotted line and the A-type repeats are filled in blue. There are two copies of this motif in the bottom group, other groups have only one copy. The top group were all samples of Zimbabwean individuals, while the bottom group were collected from individuals from different populations. Sa, Saami; J, Japanese; K, Kenyan; C, North European; Ba, Basque.

in the rest of the world for another 30 000 years. Remains of modern humans are found in China dated 67 kya, Australia 55 kya, Europe 40 kya and America 35–15 kya.

There has been considerable debate concerning the origin of modern humans. Some palaeontologists maintain that regional differences in human anatomy can be traced back to *Homo erectus* who lived over 1 mya. This has led to the **multiregional evolution hypothesis**, which states that *Homo sapiens* evolved independently from *Homo erectus* on multiple occasions in different parts of the world. This has seemed an unlikely hypothesis to geneticists because it would require long periods of parallel evolution in which exactly the same genetic changes occurred to produce the same organism in different parts of world. The alternative is the **out-of-Africa hypothesis**, in which modern humans emerged from Africa 100 kya and spread to the rest of the world, displacing archaic human populations.

The genetic evidence, as we have seen in the preceding sections, overwhelmingly favours the out-of-Africa hypothesis. One explanation for the apparent regional continuity is that there was gene flow between modern humans and archaic forms. If this happened, there is no evidence of it occurring in the mtDNA record, where it would have been expected to leave a trace. It has now been formally shown that Neanderthal man was not ancestral to anatomically modern humans. DNA from the bones of a Neanderthal man has been amplified by PCR. The sequence of a 377 bp mtDNA fragment differed from mtDNA from modern human populations by an average of 27 nucleotide changes, while in the same region the average divergence between modern human populations was eight nucleotide differences. This places the Neanderthal mtDNA sequence outside the range of variation found in modern humans. If Neanderthals were ancestral to modern humans the range of variation in the Neanderthal mtDNA pool should have been about four times greater than the variation in the mtDNA pool of modern humans. Furthermore the estimated age of the common ancestor of Neanderthals and modern humans (700 kya) is about four times greater than the estimated age of modern human populations (120 kya). This suggests that Neanderthals were not ancestral to modern humans nor did they contribute to the mtDNA of modern humans.

Combining the genetic with archaeological evidence we can build up a picture of human history. Modern humans probably originated in Africa 100–200 kya. The first modern humans are found in Israel about 100 kya, but apparently did not start to spread to the rest of the world until 60 kya. As anatomically modern humans migrated from Africa they split into two populations. One branch spread through northern Asia and on to the Americas: this branch also populated Europe. The other major branch spread through southern Asia and populated Australia and New Guinea. Polynesia was also populated from this branch, but only in recent history.

The modern humans who lived in Israel 100 kya were contemporaneous with Neanderthal populations, who continued to live in Israel for at least another 30 000 years. Moreover the last Neanderthal remains disappear from the fossil record only 30 kya. So modern and Neanderthal populations coexisted in Europe for at least 10,000 years. The replacement of Neanderthals was thus a relatively slow process and presumably occurred because modern humans enjoyed some competitive advantage. One strong possibility is that modern humans had evolved a higher level of linguistic ability.

9.3.5 Origin of genetic diversity

The effective size of the original population that emerged from Africa has been estimated from nuclear polymorphisms to consist of about 10 000 individuals and from mtDNA variation to consist of 6000 females. Very few polymorphisms are unique to one population; if a polymorphism is found in one population, it is found in all populations. What differs between populations are the allele frequencies. This suggests that most genetic variation found today worldwide was present in the original founder population of all humans. The only exception to this statement are disease alleles such as sickle cell anaemia and the CF ΔF508 allele, whose origin is discussed below.

As the founder population expanded, small groups would split off to found local populations. Each time this occurred, the founder effects outlined above would act to differentiate the genetic structure of these groups. From the time modern humans emerged in Africa to the advent of agriculture 10 kya, population numbers were low: the total world population was probably only 3–4 million. Local breeding populations would have been small, and the effect of genetic drift large. During this period the genetic variation that distinguishes different populations today would have originated. Selection would have affected alleles that influence resistance to disease, such as those at the HLA locus. Also during this time, selection would have acted on skin colour and body form in response to local climatic conditions. The advent of agriculture caused an increase in population numbers as the carrying capacity of the land was increased 10–100 fold. This increase would have frozen the effects of genetic drift. The genetic structure of present populations is thus largely a consequence of human history in the palaeolithic period when we were hunter–gatherers.

Agriculture evolved independently in three different parts of the world: Central America, South East Asia and the Middle East. In Europe at least, agriculture was spread by the expansion of farming populations into the land occupied by mesolithic hunter–gatherers. The effect of this expansion can be clearly seen in the present-day distribution of genes in Europe. Figure 9.9 shows maps plotting the spread of agriculture, the synthetic map of Europe derived from PC analysis and the distribution of the ΔF508 allele of the *CFTR* gene. The synthetic map shows a close correspondence with the spread of agriculture. This suggests that the population of **neolithic** farmers expanded from an origin in the Middle East into lands occupied by mesolithic hunter–gatherers. The more numerous farmers absorbed the smaller populations of hunter–gatherers. This is known as the wave advance model, which mathematical analysis has shown would generate the gradients in allele frequencies observed. The wave advance model has shown that the previous theory, that agriculture spread by cultural diffusion, is probably incorrect.

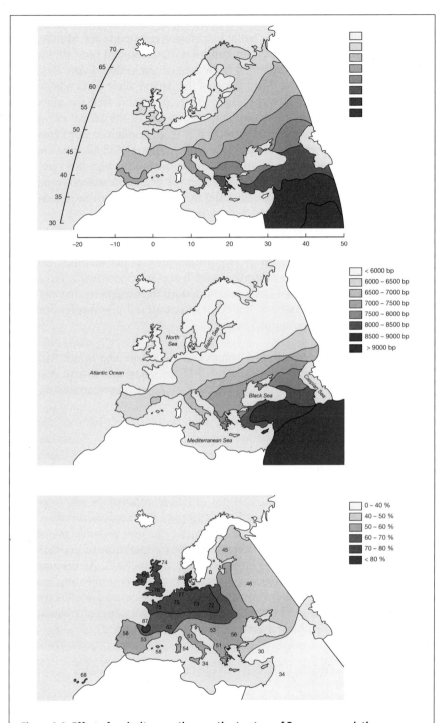

Figure 9.9 Effect of agriculture on the genetic structure of European populations.
Top: synthetic map of Europe based on values of the first PC. *Middle*: timings of the spread of agriculture based on the archaeological record. The scale is years before present. *Bottom*: Distribution of the ΔF508 allele as a percentage of the total CF alleles.

European languages, with a few exceptions, are members of the Indo-European family, which is also thought to have originated in the Middle East. We can therefore surmise that the neolithic farmers spread both their genes and their language as they expanded from their origin in the Middle East.

We can also see that the distribution of the ΔF508 allele is correlated with the genetic map and the spread of agriculture, suggesting that the allele was present in the mesolithic inhabitants of Europe before the arrival of agriculture. We shall see below that another line of evidence confirms this conclusion. Interestingly, the highest frequency of the ΔF508 allele is found in the Basque region, which has a non-Indo-European language not related to any other extant language. It is possible that the present-day Basque population is genetically and linguistically directly descended from the mesolithic population that existed before the arrival of neolithic farmers.

The preceding paragraphs describe how world populations evolved up to about 500 years ago, producing genetic diversity in local population groups. These would be mixed by migrations such as the neolithic farmers described above and other migrations we have not considered. However, starting about 500 years ago large-scale population movements have started to mix world populations on a scale not encountered previously. Some of these migrations are summarised below.

1. Europeans expanded into the Americas, Australasia and South Africa.
2. West Africans were transported to the Caribbean and North America through the slave trade. Recently, populations of Afro-Caribbeans have further migrated to the British Isles.
3. Migrations from the Indian subcontinent to the British Isles.
4. Migration from the Far East to the USA.

As a result of these migrations populations have been partially but not fully mixed, so the genetic structure of populations, such as those of the present-day USA, can be very complex. This has practical importance where genetic factors are involved in disease. For example, haemoglobinopathies are found in the Caribbean, the USA and the UK, not because malaria is endemic in these countries but because of the immigration of peoples who originated in areas where malaria is endemic (see below). As discussed in Chapter 5, population heterogeneity is a major factor complicating association studies, because an allele appears positively associated with a disease, not for a genuine biological reason, but because it has a higher frequency in ethnic groups that also have a high frequency of the disease.

9.4 Disease frequencies in different populations

We saw above that selection would normally act to restrict the population incidence of severe monogenic diseases to be close to the mutation rate. There are a number of diseases where in certain populations this is manifestly not the case. Table 9.1 lists some examples. There are two main

Table 9.1 Examples of elevated frequencies for monegenic disease in particular populations. The low frequency of type 1 diabetes (IDDM) in Japan is given as a comparison with the high level in Scandinavia.

Disease/allele	Population	Frequency/10³ population	Possible cause
Sickle cell disease	Africans	10–20	Heterozygous advantage
Thalassaemia	Africans/ Mediterraneans/ S.E. Asia	10–20	Heterozygous advantage
Cystic Fibrosis	N. Europeans	0.5	Heterozygous advantage?
Tay–Sachs disease	Ashkenazi Jews	0.17–0.4	Founder effect or heterozygous advantage?
Gaucher's disease	Ashkenazi Jews		Founder effect
BRCA1 185delAG	Ashkenazi Jews	10	Founder effect
BRCA2 617delT	Ashkenazi Jews	10	Founder effect
Porphyria	S. African (white)	3	Founder effect
MLH1 (HNPCC susceptibility)	Finnish	14 families	Founder effect
Polydactyly	Amish community (USA)		Founder effect
Type 1 diabetes	Scandinavia Japan	2 0.03	Environment and genetic effects

reasons why disease frequencies in one population may be higher than expected: heterozygous advantage and founder effects/inbreeding.

9.4.1 Heterozygous advantage

Haemoglobinopathies

Heterozygous advantage occurs when an allele that is deleterious as a homozygote is advantageous as a heterozygote. This results in a **balanced polymorphism**, where selection against the allele in the homozygous state is balanced by selection for the allele in the heterozygous state. The high frequency of haemoglobinopathies in countries where malaria is endemic is a balanced polymorphism. The geographical distribution of malaria correlates with the frequency of sickle cell anaemia and the thalassaemias. Haemoglobin S (Hb S) is responsible for sickle cell anaemia (see Chapter 4). Hb S differs from the wild-type protein by a substitution of glutamate by valine at residue 6 in the β-globin polypeptide. Hb S crystallises at low oxygen tensions and as a result the red blood cells collapse into a sickle shape. Hb S heterozygotes are clinically normal but are resistant to malaria for reasons not currently known. The allele frequency of Hb S has equalised to 0.1 in a number of different geographic locations. This exactly corresponds to the expected value calculated from the observed Darwinian fitness of the three different genotypes.

There are four African regions where the allele frequency is at a maximum: Bantu, Benin, Senegal and Saudi. All Hb S alleles are the result of the same nucleotide substitution. However, the linked haplotype is different in each region and corresponds to a haplotype that is locally common in each case. This suggests that the mutation occurred on four separate occasions, although it is possible that it occurred once and has been introduced to the four different genetic backgrounds by migration.

Heterozygous advantage is also likely to be responsible for the elevated frequencies of α and β thalassaemias in areas where malaria is endemic. β Thalassaemia is more common in Europe and Africa and less common in Asia. Malaria used to be endemic in Mediterranean countries, although it has now been eradicated. This explains the high frequency of β thalassaemia found around the Mediterranean. Indeed the word 'thalassaemia' is derived from the Greek word *thalassa*, meaning sea.

Cystic fibrosis

Among northern Europeans the incidence of CF is about 1 in 2000. On average, 70% of CF mutations are the ΔF508 allele. There is a controversial estimate that the ΔF508 mutation occurred more than 52 000 years ago (Box 9.1). The next two most frequent CF mutations also occurred on a chromosome with the same haplotype that is rare in Europe. This suggests that the common CF mutations occurred in a population that had a different genetic background from modern Europeans. The most likely scenario is that they occurred in the first palaeolithic human population in Europe, after the founder group had split off from the main stock of modern humans expanding out of Africa. Eventually, as we saw above, this population was absorbed by the expansion of neolithic farmers from the Middle East.

Why has the ΔF508 mutation maintained such a high frequency? Founder effects could have been responsible for a high frequency initially. However, the minimum age of the mutations means that selection against the homozygote would long ago have reduced the allele frequency below the present figure of 1.4%. The likely explanation is heterozygous advantage. It is possible that the reduced function of the CFTR protein in the heterozygote protects against infantile diarrhoea as it could reduce water loss across the intestinal mucosa. Recently, there has been support for this hypothesis from experiments in mice. If this is the explanation, it is interesting that more than one CF mutation occurred and was selected for in European palaeolithic populations, whereas similar mutations were not established elsewhere in the world. Perhaps this reflects differences in the environment or lifestyle of the early Europeans.

BOX 9.1: IS THE ΔF508 ALLELE ANCIENT?

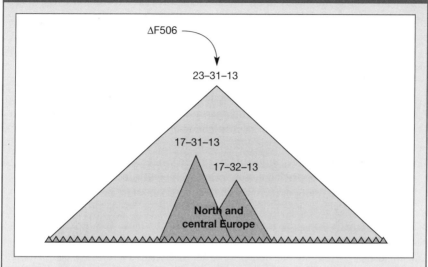

The history of the Δ508 mutation can be traced by analysis of closely linked mini-satellite markers that are in linkage disequilibrium with the mutation. Three polymorphic dinucleotide repeats have been identified in *CFTR* introns: IVS8CA, IVS17BTA and IVS17BCA. The genotypes at these loci were determined in 1738 CF chromosomes. A total of 54 different haplotypes were observed, but only three of these were common, all the other haplotypes having a frequency of less than 1%. This suggests that the mutation originally occurred in a chromosome which had one of these haplotypes; since the haplotype 23-31-13 is the most common it is assumed that this is the ancestral chromosome in which the ΔF508 mutation occurred (the haplotype is represented as the number of repeats at each locus). The other three common haplotypes could have arisen from 23-31-13 by mutation shortly afterwards (see figure). Further mutations probably gave rise to 46 of the ΔF508 haplotypes, the remaining seven arising by recombination.

A phylogenetic tree of the evolution of these haplotypes from the ancestral chromosome provides an estimate of the total number of mutations to the microsatellite loci that have occurred. From this it can be calculated that the mean number of mutations per ΔF508 chromosome is 0.867. During the collection of data for this study, 3000 meioses were studied, but no mutations were observed. This sets the maximum rate of mutation to be 3.3×10^{-4}/locus/gamete. From these figures, the number of generations that have passed since the origin of the ΔF508 mutation is calculated to be 2627. Assuming 20 years per generation, the ΔF508 mutation therefore arose 52 kya.

The haplotype 23-31-13 is rare in the general European population, which suggests that ΔF508 occurred in a population with a different genetic background than present-day Europeans. With respect to chromosomes carrying the ΔF508, 23-31-13 is evenly distributed throughout Europe. However the next most common haplotypes, 17-31-13 and 17-32-13, are more common in Central Europe (Hungary, Czech Republic, Slovakia and Germany) and Scandinavia compared with the rest of

Europe. From these observations we can reconstruct the history of the ΔF508 mutation (see figure). It was already present in the first modern humans that populated Europe 52 kya or more, on a chromosome with the 23-31-13 haplotype. However, it is not found in other world populations, so it must have arisen after the European founders split from the main body of humans as they expanded out of Africa. Subsequently, an expansion occurred in Central Europe and Scandinavia that spread the ΔF508 mutation on chromosomes with the 17-31-13 and 17-32-13 haplotypes and is responsible for the increased frequency of the ΔF508 allele in these areas today. The geographical distribution of these haplotypes and their estimated age correlates with the advance of the Linear Pottery Culture in the neolithic period. So the expansion of this culture could have been responsible.

These conclusions are highly controversial. They have been challenged on the basis of the uncertainty of the minisatellite mutation rate (see Further reading). The alternative explanation for the high frequency of the ΔF508 allele is that it is a result of genetic drift. The rare haplotype of the chromosome which carries the mutation may simply be due to the fact that the original mutation occurred on a chromosome with a rare haplotype. However, as discussed above, the distribution of CF alleles in Europe is consistent with the allele being present in palaeolithic populations absorbed by the advance of neolithic farmers. This provides independent evidence that the ΔF508 mutation is ancient.

9.4.2 Founder effects

Founder effects occur when a small group, separated from the main stock, expands in number. As a result, disease alleles present by chance in the founder population may have increased in frequency under the influence of drift and the increased inbreeding that would occur in small isolated groups. This can result in local populations with a particularly high frequency of a monogenic disorder that is rare elsewhere. Sometimes these populations originated by migration of founder groups in historical times and the identity of the individuals who carried the disease allele is known.

Ashkenazi Jews

Table 9.1 shows that a number of different diseases are more common in Ashkenazi Jews than among other groups and are maintained at a higher rate than can be explained by mutation. It has been suggested that the high rate of Tay–Sachs disease is due to carriers being more resistant to tuberculosis, which was endemic in the tenements of eastern Europe where Ashkenazi Jews lived. An alternative explanation is that the high frequency of all these diseases is explained by a founder effect. The *BRCA1* 185delAG and *BRCA2* 617delT mutations are extremely rare outside this population group and so are almost certainly founder mutations.

The Ashkenazi Jews originated in the vicinity of Strasbourg in eastern France in the early Middle Ages. They expanded eastwards at the time of the Crusades and subsequently to America in the late nineteenth and early twentieth centuries. During this time their culture and religion operated to maintain an effective reproductive isolation from neighbouring populations. From the time of their migration away from Strasbourg to the present day, their population number would have increased several thousand-fold. They thus conform to a classic founder population: selection against homozygotes would not have had sufficient time to operate to reduce the frequency of deleterious alleles. Furthermore, the *BRCA1* and *BRCA2* mutations would have limited effect on fitness since generally they have their action after child-bearing age.

Complex disease

Table 9.1 also shows that a complex disease such as IDDM is unevenly distributed in different ethnic populations. Part of this variation could be environmental in origin. However, Chapter 5 showed that the HLA alleles DR3 and DR4 contribute about 40% of the genetic risk. These alleles themselves show variation in allele frequency between populations, so susceptibility to common diseases may vary between populations because of their different genetic structures. It may be that different alleles at the HLA locus are selectively neutral and that ethnic variation reflects the effects of genetic drift. Alternatively, they have been selected for because they conferred resistance to a disease that was historically more common in certain areas.

In the case of diabetes, the situation is even more complex. The DR3/DR4 alleles do not themselves cause the susceptibility to diabetes; they are in linkage disequilibrium with the real culprits. Thus in people of northern European extraction the DR3/DR4 alleles act as markers of the chromosomes that carry the susceptibility alleles. In other populations, the susceptibility alleles show different linkage configurations and the association between DR3/DR4 and diabetes is not so marked.

9.5 Summary

- Genes that have more than one allele are said to be polymorphic. The frequency of each allele in randomly mating populations is described by the Hardy–Weinberg distribution, which states that $p^2 + 2pq + q^2 = 1$.

- Selection will increase the frequency of an advantageous allele and decrease the frequency of a deleterious allele. It is a powerful force that acts to restrict the incidence of serious genetic disease to the frequency at which new alleles arise by mutation, unless other factors are operating.

- Many alleles are selectively neutral. Their frequency will fluctuate due to sampling errors until they become fixed or extinct. The size of the fluctuations and speed at which the alleles become fixed or extinct depends upon population size.

- Migration between populations tends to reduce the effects of drift.

- Geographical distance, linguistic and cultural barriers reduce mating and define populations. However, it is often difficult to draw precise boundaries and one population usually merges gradually into the next.

- Analysis of polymorphisms of protein-coding genes allows a phylogenetic tree of human populations to be constructed. It shows that modern humans originated from a founder population of about 10 000 individuals who lived in Africa 100–200 kya. This conclusion is supported by analysis of mtDNA and minisatellite structure.

- Genetic analysis of human populations does not support concepts of innate racial differences with respect to innate abilities or characteristics.
  - The phylogenetic tree does not correlate with a tree drawn up on the basis of racial characteristics. Skin colour, facial features and body form probably evolved after the main radiation of modern humans in response to local climate.
  - Humans are a genetically uniform species, with most variation evident *within* rather than *between* populations.

- Most of the present-day genetic variation between populations probably evolved in palaeolithic times when population numbers were low. The major force determining allele frequencies was probably random genetic drift, which would have been frozen when population numbers increased after the advent of agriculture. Selection acted to alter allele frequencies that affected resistance to infectious disease.

- Genetic structures of populations were altered by migrations and expansions such as that which spread agriculture across Europe. The largest change has been in the last 500 years, which has resulted in large-scale mixing of populations especially in countries such as the USA.

- Many genetic diseases are present in certain populations at rates that greatly exceed the level at which new alleles are introduced by mutation. This can occur for two reasons: heterozygous advantage and founder effects.

- Haemoglobinopathies are present at a high level where malaria is endemic because carriers are more resistant to malaria. As a result, selection against alleles in the homozygote is balanced by selection for the allele in heterozygotes.

- CF is much more frequent in Europe than in other areas. The most frequent allele, ΔF508, was present in mesolithic populations absorbed by the advance of neolithic farmers. The divergence of intragenic microsatellite loci in linkage disequilibrium with ΔF508 suggests that the mutation is at least 52 000 years old. Since selection would long ago have reduced its frequency, it is necessary to postulate that the heterozygote has been at a selective advantage. It is possible that CF carriers were more resistant to infantile diarrhoea. The estimate of the age of the ΔF508 mutation is highly controversial and has been challenged because of the uncertainty of minisatellite mutation rates.

- Many populations founded by small groups of migrants have elevated frequencies of otherwise rare diseases.

- Ashkenazi Jews have elevated frequencies of several different diseases, which probably arose through founder effects. Some argue that Tay–Sachs disease is an example of heterozygous advantage.

- Susceptibility to complex diseases such as diabetes is also affected by population variation in allele frequencies. These may reflect variation of previously neutral alleles due to random genetic drift, or the susceptibility alleles may influence resistance to infectious diseases.

Further reading

Phylogenetic trees and nuclear polymorphisms

CAVALLI-SFORZA, L.L., MENOZZI, P. AND PIAZZA, A. (1994) *The History and Geography of Human Genes*. Princeton University Press, Princeton, New Jersey.

Contains a full exposition of the argument that all mitochondria must logically trace back to a single woman.

LI, W.H. (1997) *Molecular Evolution*. Sinauer Associates, Sunderland, Massachusetts.

NEI, M. (1996) Phylogenetic analysis in molecular evolutionary genetics. *Annual Review of Genetics*, **30**, 371–403.

Evolution of mitochondrial DNA

CANN, R.L., STONEKING, M. AND WILSON, A.C. (1987) Mitochondrial DNA and human evolution. *Nature*, **325**, 31–36.

STONEKING, M. (1996) Mitochondrial DNA and human evolution. In *Human Genome Evolution*, Jackson, M., Strachan, T. and Dover, G. (eds), Bios Scientific Publishers, Oxford.

Contains a full exposition of the argument that all mitochondria must logically trace back to a single woman.

Age of the cystic fibrosis ΔF508 mutation

KAPLAN, N.L., LEWIS, P.O. AND WEIR, B.S. (1994) Age of the ΔF508 cystic fibrosis mutation. *Nature Genetics*, **8**, 216.

MORRAL, N., BERTRANPETIT, J., ESTIVILL, X. *et al.* (1994) The origin of the major cystic fibrosis mutation (ΔF508) in European populations. *Nature Genetics*, **7**, 169–175.

MORRAL, N., BERTRANPETIT, J., ESTIVIU, X. *et al.* (1994) In reply. *Nature Genetics*, **8**, 216–217.

Neanderthal man was not our ancestor
KRINGS, M., STONE, A., SCHMITZ, R.W., KRAINITZKI, H., STONEKING, M. AND PÄÄBO, S. *et al.* (1997) Neanderthal DNA sequences and the origins of modern humans. *Cell*, **90**, 19–30.

CHAPTER TEN

DNA fingerprinting

Key topics

- DNA as a unique human identifier
- Minisatellite-based method
  - Multilocus probes
  - Single-locus probes
- Simple sequence length polymorphisms

10.1 Introduction

The human genome contains 3000 Mb of DNA; of this, selection will constrain those sequences (<5% of the total) that encode proteins. While to some extent other factors will act to conserve the sequence of the remainder, we may expect that most human DNA sequences will be highly variable. As a result the exact sequence of each individual human genome is undoubtedly unique (apart from identical twins). Chapter 3 showed how this variation has been exploited in the production of genetic maps.

The individuality of DNA sequences has practical importance in the forensic analysis of biological material left at the scene of a crime or in the resolution of paternity disputes. The DNA sequence within biological specimens, such as blood, semen or hair roots, should unambiguously identify the person from whom they were derived. By analogy to classical fingerprinting, based on the pattern of ridges on the pads of fingers and thumbs, the sequence of the DNA within a biological sample forms a DNA fingerprint that uniquely identifies the person from whom it originates. Thus it could be proved that a semen sample originated from a rape suspect or a blood-stain or hair root came from a murder suspect. In cases of disputed paternity, the children's alleles should be inherited from the parents in a simple Mendelian fashion, so a comparison of the DNA profiles of the children and claimed parents will readily determine the nature of the biological relationship.

To use the potential of DNA sequence variability, a laboratory test will need to fulfil three criteria.

1. It must be simple to carry out by technicians who may not have the level of experience and skill found in a research laboratory.
2. It must provide data that uniquely identify each individual with a clarity that will withstand critical examination in a law court.
3. It must be capable of operating with degraded DNA, since the biological material may be old and not well preserved.

In 1985 Alec Jeffreys, a British geneticist working with a minisatellite sequence located within an intron of the myoglobin gene, realised that minisatellites could be used as the basis of a **DNA fingerprinting** technique that met these criteria. This chapter describes the use of minisatellites in the original DNA fingerprinting procedure and then reviews more recent techniques designed to use the variability inherent in DNA to **profile** individual identity.

10.2 Use of minisatellites for DNA fingerprinting

We considered the general structure of minisatellites in Chapter 3. The myoglobin minisatellite consists of four repeats of a 33-bp sequence. The length of this minisatellite was not polymorphic, but in low **stringency** Southern hybridisation experiments it cross-hybridised to other minisatellites that were polymorphic. A random selection of these were cloned and sequenced. The sequence of the repeat unit in each case contained an 11–16-bp core that was identical or very similar to the sequence GGAG-GTGGGCAGGARG (R indicates a purine). In each case, the remainder of the sequences was different. The myoglobin minisatellite cross-hybridises only weakly with other minisatellites because the 17 bp of non-core sequence interferes with hybridisation. Jeffreys and his colleagues isolated two minisatellites, 33.6 and 33.15, which consisted only of tandem repeats of different versions of the core sequence. These hybridised efficiently to other minisatellites in low-stringency conditions. The pattern of hybridisation was found to be very sensitive to the actual sequence of the repeated element in the probe. Because 33.6 and 33.15 contain different sequences, they each produce a different pattern of hybridisation and hence a different fingerprint from the same individual.

Each probe produced a complex profile of about 36 bands, representing alleles drawn from a pool of about 60 minisatellite loci (Figure 10.1). The variation caused by differences in the number of repeats is maximised by removing as much flanking DNA as possible. This is done by digesting the DNA with the restriction enzyme *Hin*fI. This enzyme cleaves at a 4-bp recognition site that occurs on average every 256 bp. Because the minsatellites are repeats of sequences much shorter than this, the enzyme does not usually cut within the repeat unit (Figure 10.2). All the loci were autosomal,

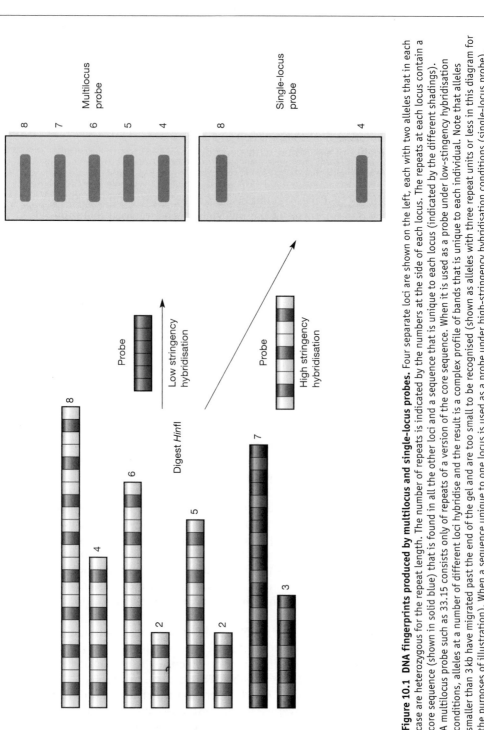

Figure 10.1 DNA fingerprints produced by multilocus and single-locus probes. Four separate loci are shown on the left, each with two alleles that in each case are heterozygous for the repeat length. The number of repeats is indicated by the numbers at the side of each locus. The repeats at each locus contain a core sequence (shown in solid blue) that is found in all the other loci and a sequence that is unique to each locus (indicated by the different shadings). A multilocus probe such as 33.15 consists only of repeats of a version of the core sequence. When it is used as a probe under low-stringency hybridisation conditions, alleles at a number of different loci hybridise and the result is a complex profile of bands that is unique to each individual. Note that alleles smaller than 3 kb have migrated past the end of the gel and are too small to be recognised (shown as alleles with three repeat units or less in this diagram for the purposes of illustration). When a sequence unique to one locus is used as a probe under high-stringency hybridisation conditions (single-locus probe), only the two alleles at that locus hybridise. The rationale for digesting with *Hinf*I is explained in Figure 10.2.

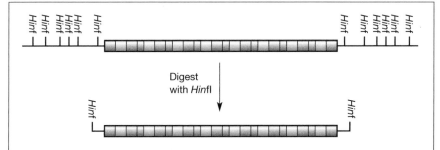

Figure 10.2 Digestion with HinfI trims away flanking DNA sequences. *Hinf*I recognises a 4-bp target sequence that occurs frequently in the sequences flanking a minisatellite. However, it does not occur in the 33-bp repeat so it will not cut anywhere in the whole tandem repeat array. Digestion with *Hinf*I therefore trims away DNA on either side of the minisatellite and so maximises the relative difference in size due to the number of repeats.

so in theory each of the two bands from a heterozygous locus could be visualised. In practice, the difference in the size of the allelic minisatellites is sufficiently great that normally only one of each allelic pair is visible within the size range resolved by electrophoresis.

The average odds that any one band is shared by two unrelated individuals is experimentally observed to be about 0.25. The chance that two unrelated individuals will share all 36 bands is thus $0.25^{36} = 10^{-22}$. Since this is considerably smaller than the reciprocal of the size of the world's population, the profile of each individual may be expected to be unique. This calculation makes the assumption that all individuals are unrelated and that the chance that each band will be shared will always be the same for all individuals. In reality, the world's population is not homogeneous and individuals belonging to the same ethnic population may be more likely to share bands than the average figure for world populations. The extent to which this biases the odds of a chance match has been highly contentious. Because of this, it is necessary to monitor the actual frequency of band sharing in different ethnic groups to validate the high degree of uniqueness claimed for the DNA profiles. This is difficult in multilocus probes because the actual loci hybridising to produce each band in the fingerprint are unknown, so it is not possible to define allele frequencies.

Even taking into account the complications of increased band sharing between members of the same ethnic group, the DNA profile or DNA fingerprint of each individual is unique. The methodology therefore meets the criteria for a simple and robust means of obtaining genetic information.

10.3 Single-locus probes

The limitation of using probes that detect multiple loci is that about 250 ng of DNA is needed for the Southern hybridisation procedure, which is often more than can be obtained from the crime scene. Moreover, since so many different loci are involved, it is not possible to use PCR to amplify small samples of

DNA. Multilocus probes are based on the conserved core sequence, while the remainder of the sequence of each repeat unit would provide a locus-specific probe. It would therefore be possible to visualise the two allelic bands of each locus, providing a much simpler autoradiogram to analyse (Figure 10.1).

A panel of probes is used that are selected to recognise highly polymorphic loci in which common alleles all fall within the range that can be analysed by electrophoresis. The probes are also characterised with respect to the frequency of band sharing between members of different ethnic groups. Since only a small number of bands are analysed a high degree of reliance is placed on each, in contrast to multilocus probes where many bands are produced. The difference between alleles that differ by one repeat unit is small in comparison to the total size range resolved by AGE. Thus it is possible that where the difference in the repeat number is small, two genuinely different alleles may not be resolved. It is therefore necessary to base the analysis on the chance that two bands would migrate to the same region of the gel. Nevertheless, it is estimated that the pattern detected by four locus-specific probes would be matched by chance less than once in a million unrelated individuals. Because single-locus probes do not produce a pattern that is guaranteed to be absolutely unique to each individual, the more neutral term **DNA profile** is now commonly used instead of DNA fingerprint.

The major advantage of locus-specific probes compared with multilocus-specific probes is that they are much more sensitive. Multilocus probes require about 250 ng of DNA, while single-locus probes require only 10 ng of DNA, which allows the analysis of DNA obtained from a blood spot or single hair root. Moreover, a defined locus can be amplified by PCR, increasing its sensitivity still further. A second advantage is that since locus-specific probes recognise both alleles of a defined locus, it is possible to measure the allele frequencies in different ethnic populations, thus providing a precise estimate of the odds that a suspect's DNA profile matches the victim's by chance.

10.4 DNA profiling based on STRs

Since 1994, STRs of the type used to construct genetic maps (see Chapter 3) have become the standard method of producing DNA profiles. The markers used are highly polymorphic and are mostly based on loci with tetranucleotide repeats, each locus commonly having between 5 and 15 alleles. If the loci are carefully chosen, a single multiplex PCR reaction amplifying three to four loci is sufficient to derive a profile. The amplified products are labelled with fluorescent dyes and analysed using an automated DNA sequencer. A number of carefully controlled tests by organisations such as the European DNA Profiling Group (EDNAP) have shown that the use of STR loci is a reliable method to obtain DNA profiles. An example is shown in Figure 10.3. Allele frequencies in different ethnic groups have been measured and a number of loci have been validated for use in forensic casework by the relevant authorities.

Compared with single-locus probes there are a number of advantages to using STR loci for profiling.

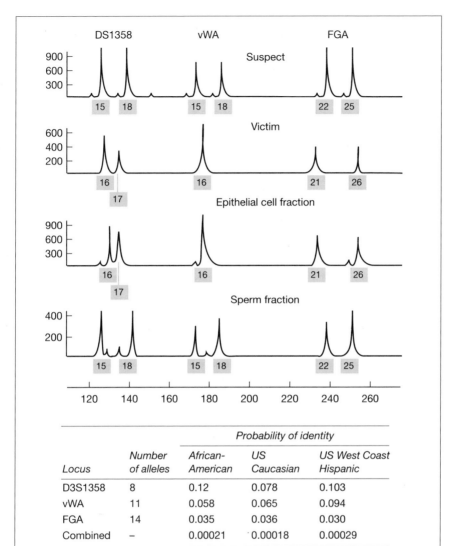

Figure 10.3 STR DNA profiling using STR loci. *Top*: readout from the ABI Prism system of four samples from a rape case. The suspect's profile matches the sperm sample. *Bottom*: ethnic distributions and probability of identity for the STR loci used.

- STR-based tests are quick; in the example shown in Figure 10.3, 32 samples were analysed in 2 hours.

- STR tests are more sensitive, requiring only 0.1 ng of DNA instead of 10 ng for single-locus probes.

- There is no need for the radioisotopes required for the Southern blots used in profiles based on single-locus probes.

- The electrophoresis conditions in the ABI Prism system are much better controlled, leading to little or no variation between lanes.

- Measurements of band migration are precise, allowing unambiguous assignment of genotype. The results can be readily stored in a computerised database.

- It is tolerant of DNA degradation in the sample. One test showed that it could be reliably used to identify victims of a mass disaster whose bodies were badly decomposed.

10.5 Summary

- The sequence of each individual's genome is unique. This provides the potential for a DNA fingerprint.

- About 60 minisatellite loci are made of repeats that have a common core sequence. If this core sequence is used as a probe in a Southern hybridisation experiment, multiple bands are seen that correspond to the alleles of the different minisatellites.

- The repeat lengths of the alleles at a minisatellite locus are polymorphic. The chance of two unrelated individuals sharing a particular band is 0.25. The chances are extremely low of sharing all the 36 bands that can normally be seen. Thus the pattern of bands is effectively unique to each individual and is called a DNA fingerprint.

- Single-locus probes are more sensitive than multilocus probes and give a simpler picture. The pattern resulting from a panel of four such probes provides a high degree of discrimination between individuals but is not necessarily unique. For this reason, it is normal to refer to DNA profiles when using single-locus probes

- There has been considerable debate about whether members of the same ethnic group are more likely to share bands and the extent to which this may erode the discrimination of DNA profiles.

- Since 1994 DNA profiles have been based on STRs. These can be analysed using automatic fluorescent-dye technology of the type used in DNA sequencing or high-density genetic maps.

Further reading

BAR, W., BRINKMANN, B., BUDLOWE, B. *et al.* (1997) DNA recommendations: further report on the DNA commission of ISFH regarding the use of short tandem repeat systems. *International Journal of Legal Medicine,* **110,** 175–176.

CLAYTON, T.M., WHITAKER, J.P., FISHER, D.L. *et al.* (1995) Further validation of a quadruplex STR typing system: a collaborative effort to identify the victims of a mass disaster. *Forensic Science International,* **76** 17–25.

DEBEBHAM, P.G. (1992) Probing identity: the changing face of DNA fingerprinting. *Trends in Biotechnology,* **10,** 96–102.

FREGEAU, C.J., and FOURNEY, R.M. (1993) DNA typing with fluorescently tagged short tandem repeats: a sensitive and accurate approach to human identification. *Biotechniques,* **15,** 100.

JEFFREYS, A.J. (1987) Highly variable minisatellites and DNA fingerprints. *Transactions of the Biochemical Society,* **15,** 309–316.

LINCOLN, P.J. (1997) Criticisms and concerns regarding DNA profiling. *Forensic Science International,* **88,** 23–31.

http://www2.perkin-elmer.com.80/fo/773201/773201.html The Perkin-Elmer Web page gives details of its commercial testing kit used to generate the data shown in Figure 10.3.

CHAPTER ELEVEN

Human genetics and society

Key topics

- Potential dangers of genetics to human society
- Gene testing
  - Prenatal testing
  - Neonatal screening
  - Diagnosis of genetic disease in children after birth
  - Presymptomatic testing for late-onset diseases
  - Presymptomatic testing for predisposition to complex diseases
- Human rights
  - Insurance and employment
  - Privacy and ownership of genetic information
- Patents
- Gene therapy
- Genetic determinism and individual responsibility

11.1 Introduction

This book commenced with a consideration of the influence that our genetic constitution has over our health. We discovered that single-gene disorders were responsible for a large part of serious childhood ill health, resulting in reduced life-expectancy and quality of life for those affected and immense suffering, distress and sorrow both to sufferers and to their families and carers. Furthermore, there are few diseases of later life not influenced in some part by genetic factors. The main part of this book has been concerned with an epic period of scientific advance where, in a few years, we have equipped ourselves with the technology to understand the molecular defects that cause these diseases, to devise efficient means for their diagnosis and to realistically contemplate novel ways to prevent or treat them. Over the next few decades this may well revolutionise medicine and fundamentally alter the way we think of ourselves.

At the same time, these advances in human genetics will have a profound impact on society; therefore, as a society, we must consider carefully how we will respond, ensuring that we use our new-found knowledge to relieve suffering and promote good health whilst avoiding the very real possibility that it could have negative consequences. This chapter explores the nature of these ethical, legal and social issues. It is important to realise that many of these issues are not in any way abstract or far off. They affect us either now or in the very near future.

There have been a number of formal investigations into these problems by groups comprising scientists, medical practitioners and moral philosophers. In the UK, the Nuffield Council on Bioethics has produced a report on the ethics of gene testing and the House of Commons Select Committee on Science and Technology has investigated human genetics (see Further reading). As a result of the House of Commons report, the Human Genetic Advisory Commission (HGAC) was established in 1997 to advise the Government on general issues relating to human genetics. In addition, the Gene Therapy Advisory Committee (GTAC) and the Advisory Committee on Gene Testing (ACGT) report to the Secretary of State for Health. In the USA, the NIH and DOE have established a Joint Working Group on Ethical, Legal and Social Issues (known as ELSI) and have set aside approximately 5% of the funds of the Human Genome Project for research into ELSI. The working group established a task force to investigate gene testing. The report of the task force is available over the Web (see Further reading). For the most part, in the USA ethical issues in genetic research and gene testing are regulated by local ethical committees.

Deliberations in this area are coloured by the historically bad record that the discipline of genetics has had in the way it has advised and influenced society. In the early part of this century, there was widespread concern that the genetic fitness of western societies was being eroded by the differential reproduction of what were labelled as 'feeble minded' or 'morally defective' classes. The answer to this was seen to be the improvement of the genetic stock by selective breeding, an approach called **eugenics**. We now appreciate that such philosophies are without any scientific justification and are universally regarded with abhorrence. Changes in the human gene pool can only take place over many generations in response to selective pressures such as plagues, endemic infections and climatic conditions.

It is difficult to appreciate now the extent to which eugenics held sway among both the scientific and political classes. Several academic journals of genetics included the word 'eugenics' in the title. Quotations from many renowned statesmen and scientists show that belief in eugenics was considered normal or even responsible for leading members of society. It was not just in Nazi Germany that eugenics was practised. In the 1920s and 1930s a number of states in the USA passed laws for the forcible sterilisation of the 'feeble minded' and tens of thousands of such operations were actually performed. Recently it has been revealed that Sweden carried out over 60 000 sterilisations for eugenic reasons, ceasing only in the 1970s. The eugenics episode serves as an illustration that the ill-advised application of genetics can be exceedingly harmful and emphasises the importance of careful deliberation about the possible consequences of recent advances.

11.2 Genetic testing

Genetic testing refers to any procedure designed to ascertain the status of any aspect of someone's genotype for purposes of diagnosing or predicting disease wholly or partly caused by hereditary factors. This covers tests based on DNA, RNA or chromosomes but may be extended to include protein-based tests or even more broadly based medical examinations. Genetic tests may be carried out in a variety of contexts:

- prenatal testing;
- neonatal screening;
- diagnosis of genetic disease in children after birth;
- screening prospective parents for the carrier status of genetic disorders common in a particular population;
- presymptomatic testing for a serious late-onset disorder;
- presymptomatic testing for genetic predisposition to a complex disease.

Before we consider these different situations in detail, there are some general points that can be made.

1. It is essential that the tests are reliable and the diagnosis clear. This is not as straightforward as it may at first appear. Some genetic tests use linked markers rather than following the mutation itself. Such tests may be falsified by recombination between the marker and the mutation. In addition, many tests look for common mutations and may fail to detect rarer ones. For example, it is straightforward to detect the *CFTR* ΔF508 mutation (see Chapter 7), but there are hundreds of rarer mutations that might affect the gene. Thus a person who may appear to be a ΔF508 heterozygote may in fact be homozygous for inactivating mutations in the *CFTR* gene. As new tests are developed, there must be a mechanism to ensure that they are subject to independent review so that their effectiveness is known to doctors and genetic counsellors who will order their use. There is also a need for quality assurance in laboratories that carry out the tests. They will be marketed and carried out by private companies for commercial gain, and we may expect a large number to appear over the coming years. Ensuring quality and reliability and preventing exaggerated claims are major priorities.

2. Experimentation on human subjects must only be carried out with the informed consent of the person being tested. This is an obligation

enshrined in international law, known as the Nuremberg Code. It was established after evidence in the Nuremberg trials after World War II in response to revelations of medical experimentation in the concentration camps of Nazi Germany. There is a distinction between research and an established medical procedure. Once the validity of a genetic test has been established by research and it is used to diagnose or counsel individual patients, it is no longer experimental and is not covered by the Nuremberg Code. Nevertheless, the consensus is that the requirement for informed consent should still apply when genetic tests are used in a clinical context. Informed consent requires that the subject fully understands the nature of the test and the consequences that may arise from it. This leads to a general need for counselling, both before the test is undertaken and to explain the results when they are available. In some cases, such as a young child or where there is mental impairment, the person most affected may not be in a position to understand the reasons for testing. There is thus the need for the parents or guardian or some other representative to be able to give consent on their behalf. This consent should only be given for immediate medical reasons. Determining the carrier status of a young girl for a sex-linked disease would be an example where this condition is not met. Children's views can be taken into account, and as they get older and their understanding grows their wishes should be given increasing importance.

3. Tests should only be carried out when there is some positive action that can be taken as a result. We shall see that these tests can sometimes result in significant detriment to the person tested. Therefore tests can only be justified if they allow the prevention or better treatment of a disease or provide some other genuine benefit.

4. Everybody has the right to know about information that affects their health and has the right to expect that this information will be kept confidential. Thus access to genetic information held on any sort of file must be secure and at the same time made available to the individual concerned. Alongside the right to know, is the right not to know. This not only underlines the need for informed consent to any test but also emphasises the need for vigilance that unwelcome information is not inadvertently obtained or passed on to the individual concerned.

5. Genetic tests have implications for the whole family and may extend beyond the immediate nuclear family. Some instances, discussed below, may lead to situations where one person's right to confidentiality conflicts with another's right to know.

Prenatal testing

Prenatal testing is carried out where there is a known risk of a serious monogenic disorder or in older women where there is an increased risk of Down's syndrome and other chromosome abnormalities. This would normally be done in order to provide the opportunity of termination in the event of an adverse result. Such testing empowers parents, who may not otherwise risk having a seriously affected child, to reproduce. For example, where they have already had one affected child, the woman can become pregnant in the knowledge that the pregnancy can be terminated if the embryo is affected. The positive value of such tests is clear. However there are a number of issues connected with embryo testing.

Prenatal tests can be invasive and carry a certain risk of miscarriage (1% for amniocentesis). The potential benefit gained from the test must be commensurate with this risk. From this it follows that tests with a significant risk should only be carried out where the condition investigated is serious and the possibility that the embryo is affected is significant.

If prenatal tests are made with a view to termination of pregnancy, it is important that the parents fully understand this. There is little point and potentially much harm in testing an embryo if the parents would not wish to terminate the pregnancy whatever the result. This means that the parents should undergo counselling before the tests are carried out. Of course, parents retain the right to proceed with the pregnancy in the event of an unfavourable outcome to the test. Because prenatal testing implies the possibility of a termination, there may be an assumption that members of certain ethnic or religious groups would not wish such testing because of an objection in principle to such procedures. It is for the individual concerned to make such a decision and it is important that the option is always presented.

A number of tests are made routinely in the antenatal clinic. It is possible, even likely, that in such circumstances the prospective mother will not fully understand the nature of the tests and the possible consequences. She may then be unprepared for the unwelcome knowledge from an adverse outcome. This would be particularly serious if she objected in principle to termination and thus would be forced to complete the pregnancy in the knowledge that the child would be seriously disabled. At the very least, she may have wanted the opportunity to decide whether she wanted the test before it was carried out. This is not to say that prenatal tests should not be carried out, just that it is important that time is taken to explain their nature to the mother.

Prenatal testing with a view to termination makes a value judgement about the worth of the unborn child. This has two important consequences. Firstly, if it is agreed that certain conditions are sufficiently serious to warrant termination, then there is a danger that this could be construed as making a judgement about the worth of people who have been born who suffer from the disease. Secondly, since there is a range of disablement that results from genetic disorders, how do we decide what constitutes a sufficiently serious disablement to warrant termination? Most people, but by no means all, agree that a serious condition that will result in a short lifespan with much suffering are sufficient grounds for termination. Many parents would not wish to have a child affected by Down's syndrome. However, parents with such children often enjoy a rich and loving relationship with them. The Down's Syndrome Association in the UK do not consider it as grounds for termination. This seems to be a decision that should properly be left to the parents. How, though, do we react to a diagnosis of a late-acting autosomal dominant disorder where the unborn child can expect to enjoy 40-50 years of normal life? The consensus here is that this is not grounds for termination. If this is the case, it becomes improper to apply the test in the first place, because self-evidently an embryo cannot give informed consent but will have to live many years with the consequences.

The decision to terminate a pregnancy as a result of a genetic test must always be taken by the parents. It is important that the parents do not come under any form of pressure. One such pressure may come from the physician treating them. Another pressure may be the financial consequences of the healthcare needed by the disabled child when it is born. Healthcare is financed in different ways in different countries. In some, such as the UK, it is financed by the National Health Service; in others, such as the USA, it is financed by an insurance-based system. Whatever the system, will it consider the ill health of a child born in such circumstances to be caused by the actions of the parents and this restrict healthcare?

Neonatal screening
Neonatal screening is carried out for the occurrence of a common genetic disorder. Screening for phenylketonuria is already widespread, as is screening for haemoglobin disorders in communities where they are endemic. Screening for CF is starting. The benefit of screening is that early detection leads to early treatment, which would prevent the onset of the disease in the case of phenylketonuria, or can greatly improve the prognosis in the case of CF and the haemoglobin disorders. Like all testing, informed consent is ideal; however such tests are routine and in practice it is unlikely that sufficient time can be devoted to any form of counselling. While this is regrettable, the benefits of such programmes are generally thought to outweigh this problem.

Carrier screening can be very effective in reducing the population incidence of a disorder. This may come about by avoiding marriage between carriers, a decision of two carriers not to have children or by using prenatal testing and termination of affected embryos. Screening for β thalassaemia in Cyprus and Sardinia has led to a marked reduction in the number of cases. Carrier screening is also carried out for Tay–Sachs disease in a number of Ashkenazi Jewish communities. In all these cases, the programme is actively supported by the relevant religious leaders. In some cases, the local religious leader may act as a clearing house by receiving the results of the tests and only acting where both partners are carriers.

The major problem with carrier screening is that it may lead to discrimination against those diagnosed as carriers. This may be in the form of social stigmatisation in finding marriage partners or lead to difficulties in obtaining insurance or employment. A compulsory screening programme for sickle cell anaemia in several American states in the 1970s is widely held to have been disastrous because it led to discrimination against carriers. A common problem in population screening for recessive disorders is a failure to realise that carrier status does not affect the health of the person being tested but will affect the health of any children resulting from union with a partner who is also a carrier. Thus the discrimination against sickle cell carriers was not only unjust but was without any basis in reality.

The need for a clear understanding of the consequences of carrier status means that counselling must be provided to those who are carriers. Since population screening will only be undertaken for common disorders, it is likely that the frequency of carriers will be high, up to 1 in 20 for CF in Europeans. This causes a problem of resources. One solution that has been suggested is to inform people of their carrier status only when their partner is a carrier as well, because only then is there actually a problem. However, in this scenario most carriers will continue to be unaware of their status and, because they have been tested without any adverse result, may assume that they are not carriers. This could cause difficulties if they subsequently had children with another partner who was a carrier.

Diagnosis of genetic disease in children after birth

Genetic tests may be carried out in a child, less commonly in an adult, to confirm a diagnosis of a genetic disease or in some cases to exclude it. Such use does not differ in principle from any other test used to aid medical diagnosis and does not present any ethical problems.

Presymptomatic testing for late-onset diseases

Presymptomatic testing may be carried out for serious late-onset diseases, such as HD, familial Alzheimer's disease, familial colon cancers or familial breast cancer. It is important to distinguish these sorts of tests from novel tests that will become available for predisposition to common diseases, which are considered separately below. For serious late-onset diseases, the presence of the mutation is highly likely to lead to the onset of the disease. However, this is not necessarily inevitable. The penetrance of *BRCA1* is only 50% by age 50 and 85% throughout life. Thus some women may have the mutation but not suffer the disorder. Equally, even if a woman does not carry the *BRCA1* mutation there is still a significant lifetime risk that she will develop breast cancer from other causes. The situation is further compounded by current uncertainty about the penetrance of the *BRCA2* mutation (discussed in Chapter 4). Even where the eventual onset of the disease is virtually certain, as is the case with HD, there is considerable uncertainty about the age of onset and the severity of symptoms.

These tests are unusual because they predict future disease in people who are currently healthy. There can be a number of real and serious consequences from an adverse result. Firstly, such a result can inflict severe distress and psychological damage on a person arising from the knowledge that at some point in the future he/she is likely to suffer from a devastating disorder. Even members of an affected family who test negative can suffer psychologically from feelings of guilt that they have escaped. Secondly, affected people will find it more difficult to obtain insurance and employment. As we shall see below, most insurance companies require the results of any genetic tests to be declared.

Because of the possibility of harmful consequences, there have to be clear positive reasons for undertaking such tests. Such reasons do exist. Firstly, there may be some clear action that can be taken to prevent or ameliorate the disorder. In the case of the colon cancers, removal of the colon has been shown to be an effective prophylactic treatment. Here the reason for undertaking the test is obvious. The situation in familial breast cancer is less clear. BRCA1 and, to a lesser extent, BRCA2 predispose to ovarian cancer at a high frequency and other cancers at a lower frequency. At the very least, prophylactic surgery would require oophorectomy (removal of the ovaries) as well as radical mastectomy. Even then we still do not know whether this will be effective: it is difficult to completely remove all breast tissue and there is still the risk of other cancers. Furthermore, it is the nature of this disease that we are unlikely to know for some years whether prophylactic surgery is effective. In other diseases, such as HD and familial Alzheimer's disease, there is no treatment and it is difficult to see any treatment becoming available in the foreseeable future.

A second advantage from foreknowledge of disease is that it may allow appropriate life choices to be made. This may include decisions about marriage and having children, financial decisions and career choices. People who know they are at risk may make very different decisions if they know for certain whether or not they are affected. One choice that is special to breast cancer is that a woman may choose to have her children at an early age, followed by prophylactic surgery.

In this context of life choices another ethical dilemma arises. These choices are only open to someone who is relatively young and who may therefore decide to be tested. They may have a middle-aged parent who is at risk but who has not yet developed any symptoms. Since there is no positive value to the parent of knowing their status, they may well decide not to be tested. The problem is that if the child is diagnosed as having the disease, this diagnoses the parent as well. How is a balance to be struck between the right of the child to know and the right of the parent not to know?

Presymptomatic testing for predisposition to complex diseases

Presymptomatic testing for predisposition to common complex diseases is not yet widely available but is likely to become available in the next few years. We have seen in Chapter 5 that type 1 diabetes has a strong genetic component and that these components are being successfully identified. However, type 1 diabetes also has a strong environmental component, since the concordance between identical twins is only 30%. It is not difficult to envisage a situation soon where neonatal screening could identify children at genetic risk, whose parents could then be advised on how to avoid environmental risk factors. Alternatively, children could be intensively monitored for early markers of the disease, such as autoantibodies to β cells. This could then lead to early and aggressive treatment to prevent the full-blown disease from developing.

Research is also well developed into genetic factors that predispose to heart disease. One dominantly acting mutation, affecting about 1 in 500 of the population, has been shown to cause familial hypercholesterolaemia. Untreated this is likely to lead to heart disease by the age of 40–50 years. However, this can be prevented by changes in diet and lifestyle or by lipid-lowering drugs. A large number of other alleles have also been identified that alter the risk of heart disease but which can be modified by drugs or by diet and other lifestyle changes.

An important difference between this type of test and tests for late-acting autosomal dominants is that the latter carry either a very high risk or virtual certainty that a disease will occur. In contrast, predisposition indicates an elevated risk but no objective certainty about the outcome. This distinction is very important. Apart from this, many of the issues that were considered above for late-acting dominant disorders apply to predisposition as well. Diagnosis of risk factors comes with potential negatives, such as psychological harm and discrimination in insurance and employment. Normally there should be a clear positive reason for undertaking the test. However, most of these tests are only likely to become widespread if they are coupled to an effective course of action for the prevention of the disease or intensive medical surveillance leading to early and more successful treatment.

The complexity of the balance sheet between the benefits and potential harm arising out of both types of presymptomatic testing underscores the need for effective counselling before the tests are carried out and to explain the implications of the results once they are available. Several research studies have shown that the proportion of at-risk individuals who wish to undertake such tests is reduced following counselling.

The difficulty of providing effective counselling is one reason why there is considerable concern at the prospect of private mail-order testing. The incentive to use such a service is that the result may be kept more confidential than would be possible if it was arranged through a doctor. The desire for confidentiality comes from the penalties that may accrue in respect of insurance and employment.

11.3 Human rights

Insurance

Life insurance pays a benefit to the dependants of the insured upon their death. In health insurance the benefit pays the costs of medical treatment in the event of the insured suffering ill health. The size of the premium that the insured pays depends on the risk of the insured event happening. Over the years, insurance companies have built up great expertise in calculating the size of these risks, based on information such as the insured person's age, occupation and lifestyle risks such as smoking. The system cannot work if the insured person knows their personal risk is higher than the insurance company realises. A simple example would be someone who tests positive for HD insuring their life for an abnormally large sum without declaring the result of the test. For the insured, this is betting on a certainty; for the insurance company, this is a certain loss which, if it happens on a large scale, will inevitably lead to higher premiums for all.

To prevent such fraud, insurance companies could demand to know the result of any genetic tests that have been carried out. This would be a disincentive to take the test in the first place. The undesirability of this outcome may be seen from consideration of a test for hypercholesterolaemia. People who test positive for this condition may save their lives by appropriate treatment and lifestyle change, but they will not be insurable. In other words, they may take the test and live or not take the test and be insured, but they are unable both to live and be insured at the same time.

As more tests become available, it may be increasingly possible to specify the likely future health prospects of everyone. There is a danger that insurance companies will only wish to insure those whose genetic health is good and discriminate against less fortunate individuals. Such a situation would be particularly serious in those countries such as the USA that rely on health insurance to fund medical treatment. In the USA, 15 states have enacted gene privacy legislation and similar legislation is pending in 30 other states (as of April 1997). Moreover, a federal bill (Health Insurance Portability and Accountability Act 1996) severely limits the scope of insurance companies to make use of gene testing for health insurance. In 1997, a similar prohibition was incorporated into the Council of Europe Convention on Human Rights and Biomedicine, although this awaits ratification by member states. While prohibition prevents discrimination on genetic grounds, it does not solve the problem of fraud. If a solution is not found to this problem, the insurance companies predict a large rise in premiums. One solution that has been proposed is to limit the pay-out in the event of death from a prescribed list of genetic diseases.

Employment
An employer may have an interest in genetic tests of prospective or current employees for a number of reasons. Firstly, employers may not wish to invest resources and time in training someone who in the future will be unable to perform their job because of ill health. Ability to perform a job at the present time and not at some point in the future is an implicit assumption of employment laws in most countries. For example, it is not permissible to discriminate against a woman on the grounds that she may become pregnant at some point in the future. It thus seems fairly straightforward to prohibit genetic discrimination on these grounds.

A second reason an employer may wish to apply a genetic test is for safety or other reasons of public interest. An often-quoted example is that of an airline pilot predisposed to sudden heart disease. It is not difficult to imagine other similar examples. When public safety is an issue it may be necessary to override the rights of the individual. However, it is important that if other means are available then they should be used. For example, an employee doing a job that required high-quality eyesight can be checked by simple eyesight tests rather than a genetic test.

Thirdly, an employer may wish to test for sensitivity to some environmental hazard such as a carcinogen, since there is clear evidence that sensitivity to xenobiotics (chemical compounds not found normally in the body) is influenced by genetic factors. We would expect that every practical method would be used to reduce exposure in the first place. Nevertheless, there may be some industries where this is not possible. It would surely be sensible to screen out those who may be particularly sensitive. If this is so, we would then face the problem of how to react to employees who refused such tests. Clearly it would be wrong to force someone to take a test against their will. However, an employer may also wish to be protected against the consequences if the employee who refused the test was to consequently suffer the ill health that could have been avoided.

Confidentiality

Everyone has the right to keep information about their genetic constitution private. There will be many occasions when this right to privacy may come under attack. For example, we have already seen in recent years sensational treatment in the media concerning the HIV status of well-known people. There is far more scope for this sort of problem in genetic testing, e.g. the future health prospects of a politician. Clearly, genetic information will need some form of legal protection

A different problem arises out of the fact that genetic testing is concerned with families as much as individuals. When one member of a family is diagnosed as being a carrier of a genetic disease, it becomes possible that other members of a family are as well. Consider, for example, a woman who has a boy affected by a sex-linked disease and it is confirmed that she is indeed a carrier and that this was not a mutation that arose during gametogenesis in her mother. There will be a 50% possibility that her sister will be a carrier as well. Suppose the sister is about to become pregnant. She clearly needs to be aware of this information so she can ascertain her carrier status. With this knowledge she can make her own decisions about her reproductive choices. She may decide not to have children, or she may go ahead and become pregnant but use prenatal testing to ensure the embryo is healthy. What happens if the first sister refuses permission for the information to be passed on? Here we have a clear conflict between the right of the second sister to know and the right of the first sister to confidentiality. We may hope that counselling will persuade the first sister to divulge the knowledge, but this does not always occur in practice.

Individual responsibility

Research into personality disorders may well lead to the conclusion that certain antisocial actions, such as substance abuse or violent behaviour, are influenced by genetic factors. There has already been a demonstration that a history of violence and criminal activity in an extended Dutch family is due to a single mutation in the monoamine oxidase gene. This leads to two contrasting ethical problems. On the one hand, individuals predisposed to such antisocial behaviour may seek to absolve themselves of responsibility for their actions. Genetic predisposition to criminal behaviour has already been used as a defence in a murder trial in the USA. An opposite problem is that society may seek to stigmatise those found to carry such traits. Since such behaviour may be more common in certain sections of the community and in certain ethnic groups, it is but a short step to label those groups as genetically inferior. The scientific fallacies inherent in this form of genetic determinism are discussed below. It is sufficient to note here that such an approach ignores the overwhelming influence of the environment and poverty on violence and drug addiction.

Another area in which individual responsibility will become important is how we react to warnings that we may be genetically predisposed to a complex disease. For example, it is not difficult to imagine in the future that neonatal screening may be carried out for predisposition to type 1 diabetes. Research may also have identified environmental risk factors that trigger the disease or have led to medical treatments that prevent its onset. Will we hold parents responsible for ensuring that a child avoids the environmental risks by keeping to a diet or undergoes suitable medical treatment? Will a disease be judged self-inflicted if people ignore warnings that they are susceptible? There is a growing trend to restrict treatment where disease is self-inflicted as a result of cigarette smoking or substance abuse. We may note here that the justice of such practices may become even more questionable if it is shown that some individuals are more susceptible to addiction than others.

11.4 Patents

Much of the current impetus for the advances in human molecular genetics comes from the private investment of the biotechnological and pharmaceutical industries. This investment is only possible if any invention of commercial value can be protected by patent, which prevents others from using the invention for commercial purposes without the permission of the patent holder. This permission is given in the form of a licence, which usually involves the payment of a royalty. In return for the issue of a patent the inventor must disclose full details of the invention. In this way society gains the benefit of the knowledge generated by the research, while the rights for the inventor to commercially exploit the invention are protected.

Each country has its own particular patent system. Typically an inventor files for a patent by describing details of the invention. This is then examined by a patent examiner, who decides if the invention meets the following criteria.

1. The invention must be novel. This means that the patent is awarded to the person who first makes the invention. Furthermore, any publication made before the patent application means that the patent is no longer novel (in the USA there is a year's grace after first publication).
2. The invention must not be obvious.
3. The invention must have utility. Essentially this means that it has a commercial application.
4. The European Patent Convention prohibits the issue of patents to inventions that offend morality or public order, for example a letter bomb or an instrument of torture.

If the patent examiner is satisfied, the patent is granted. However, it can still be challenged in a court of law.

It is important to understand the difference between an invention and a discovery, because this difference is the basis of a fierce controversy concerning the application of patents to human genetics. An invention is a way of doing or making something better than it has been done before: it depends on human ingenuity. A discovery relates to something that already exists in nature. Its existence owes nothing to human ingenuity, although much ingenuity may have gone into making the discovery.

Although patent laws in different countries are harmonised, there are significant differences. Most European countries subscribe to the European Patent Convention (EPC). Patents may be applied for in individual countries or in the central office in Munich. Taking out patents in individual countries is cheaper but only provides protection in that country. In the USA, an invention may be discussed publicly for a year before filing but in Europe an invention may not be revealed publicly before the patent application. As a result, there is a difference between the method used for deciding priority: in the USA priority goes to whoever was the first to invent; in Europe it is the first to file. The importance of demonstrating priority in the USA has led to the practice of notarised laboratory notebooks in commercially funded research.

There is general agreement that it is right that commercial companies should be able to patent in the field of human genetics. Where there is controversy is whether patents should be issued in respect of cloned genes and their nucleotide sequences. Four objections are commonly made.

1. It is ethically wrong for any one person to claim intellectual ownership over any piece of information contained within the human genome. In essence, this objection relates to a part of European patent law that prohibits the patenting of parts of the human body.
2. Cloning a gene and determining its nucleotide sequence is a discovery not an invention. An inventive step of commercial utility may follow the discovery, for example a diagnostic test for a common mutation. This could be patented but not the gene itself.
3. The person who holds the patent controls any further commercial process that depends upon that sequence. This has two important consequences. Firstly, the patent holder may not use the information for any commercial application, but may prevent any others from doing so. Society is thus denied the benefit of a treatment or diagnostic test that may be derived from the gene. There is a rarely used provision to compel the patent holder to grant a licence that could be used in this situation. Secondly, when a patent is granted it is usually open to others to develop different and better ways of doing the same thing: one may patent a mousetrap, but this does not stop others from patenting a better mousetrap. This cannot apply in the case of any process that depends upon a gene which has been patented. The patent holder prevents any alternative method from being developed. Indeed the interest

of the patent holder at the time of patenting may be quite different from subsequent uses of the gene. There may be uses unimaginable at the time of patenting, but the original holder still controls this application.

4. The essence of human genetic research is that it is cooperative. The process of cloning a disease gene may start with individual family members collecting information, which may be assembled into family trees by local research laboratories and physicians. Mapping and cloning the gene requires the use of dense maps and probably nucleotide sequence data obtained from publicly accessible databases. Both the maps and the databases were assembled with great effort by many different laboratories. It is often remarked that genes seem to be independently cloned by several groups within a few weeks. There is a reason for this: once the information has been collected that allows the gene to be located, there are many competent laboratories that can make the last step. Granting a patent to any one of them provides a reward to one party for the collective labours of many others.

The controversy over patenting DNA sequences came to the fore in the USA with an application from the NIH to patent ESTs for which no function had been established. This patent was not allowed because it was deemed that because their function was not known, the sequences could not have utility. However, provided there is a use that can be demonstrated, the patenting of cloned genes whose function is known, together with their nucleotide sequence, has been allowed in the USA, and after considerable debate is likely to be allowed in European patent law. The overriding factor in bringing this about was the unanimous view of the biotechnology/pharmaceutical industry that they could not operate without it.

11.5 Gene therapy

Gene therapy promises for the first time to cure a genetic illness rather than ameliorate the symptoms. It may be carried out using two fundamentally different strategies. In somatic gene therapy, the genetic constitution of a person is modified or supplemented in such a way that the change is limited to that person's own body and is not passed on to the next generation. In germline therapy, a person's germ cells, or cells that will become germ cells, are changed so that the change is passed on to the next generation.

Somatic gene therapy is theoretically no different from any other form of medical treatment that seeks to rectify a physiological dysfunction. The view that it is similar to an organ transplant is not entirely accurate because gene therapy requires no donor other than the cells from which the original DNA was cloned. Provided that the gene therapy is directed towards treating a disease, rather than enhancing natural characteristics, somatic gene therapy requires no further control than would normally be applied to any experimentation involving human subjects, the ethical principles of which are well established. If the research is successful, then marketing of gene therapy products will be subject to the same controls as any new drug, i.e. clinical trials will be needed to prove that it is both safe and effective.

Germline gene therapy in principle offers many advantages over somatic gene therapy. Once a reliable method is found of manipulating the germline then that method would be generic for all genes, whereas somatic therapy relies on different *ad hoc* methods of introducing genes to different tissues according to which is affected by a particular mutation. For example, somatic gene therapy for CF requires genes to be delivered to cells lining the airways of the lung. Even if this is successful, other affected cell types such as those of the vas deferens or the pancreatic duct are not treated. Somatic gene therapy for DMD requires genes to be delivered to muscle cells and for HD would require genes to be delivered to brain cells. Each disease presents its own special problems and will require its own research programme to solve them.

Germline therapy would make it much simpler for a couple at risk of having a child affected by a genetic disease to have normal children. At present, each pregnancy must be tested and affected embryos aborted. Sometimes the laws of chance operate unkindly and it is not unknown for a series of pregnancies to have an unfavourable outcome. It has been argued that it is better to fix the problem before conception than to terminate the life of an embryo. In this context it should be noted that preimplantation testing (see Chapter 7) may provide a more realistic way of solving this problem than germline gene therapy.

There are considerable practical difficulties to be overcome before germline gene therapy can be contemplated. Somatic gene therapy aims to supplement the defective gene by introducing the functioning gene in such a way that it does not integrate into the genome. Germline therapy will require the replacement of the defective gene with the functioning copy. If the new gene is not precisely targeted to the location of the defective gene, two deleterious consequences may follow. Firstly, the original mutation may reappear in subsequent generations due to reassortment of the genetic material at meiosis. Secondly, insertion of new DNA at another site could be mutagenic: it may either inactivate another gene with unforeseeable consequences or change the pattern of expression of another gene perhaps with oncogenic consequences. Techniques are not currently available for this type of targeting. A second technical difficulty is the current lack of any method for delivering genes to the germline. These difficulties are likely to prevent germline gene therapy for some years. Nevertheless, we may reasonably expect these problems to be solved at some time in the future.

There are multiple ethical objections to germline gene therapy. Firstly, it is perceived by many to be 'playing God' by interfering with the basic genetic constitution of the human species. Secondly, it opens up the road to manipulation for eugenic reasons, i.e. manipulation designed to enhance characteristics such as intelligence, musical or sporting ability, etc. Thirdly, it is irreversible; if for some reason a change turns out to be in error it cannot be removed from the human gene pool. Fourthly, its development will require a large amount of experimentation on human embryos and, at some point, will lead to experimental humans who will be at considerable risk of suffering from a genetic dysfunction. By definition, such humans will be unable to give their prior informed consent.

The combined practical and ethical objections have meant that germline therapy is prohibited in all countries that have the research capability to attempt it. Nevertheless there are many scientists, genetic interest groups and even religious leaders who are of the opinion that the benefits outweigh the drawbacks. Should the technical problems be solved the controversy is likely to resurface.

11.6 Genetic determinism

One of the themes that runs throughout this book is that genetic factors influence the risk of common diseases and psychiatric disorders. It is also highly probable that our personalities and abilities are also influenced by genetic factors. We have seen that the means have been developed to identify these genetic factors and that we may expect that over the next few years these techniques will be applied to the analysis of the causes of many common diseases and psychiatric disorders. Inevitably they will also be applied to an attempt to improve our understanding of the origin of fundamental human characteristics.

In doing this, there is a danger that we will come to regard our health, personalities and abilities to be mechanically determined by our genetic constitution. Such a view is scientifically wrong. It is important to be fastidious about the relationship between genotype and phenotype because it is easily caricatured by both sides of the political spectrum. Susceptibility to common diseases and the development of phenotypic traits are likely in each particular case to be controlled by several genes that interact with each other and with the environment. These interactions will be complex and are unlikely ever to yield simple predictions about the outcome. Moreover, because the contribution of the environment is nearly always significant, the outcome will usually be modifiable by environmental change.

It is worth considering what we know about the genetic predisposition to IDDM because it is instructive in these matters. Recall that the concordance in identical twins is 30%. Immediately, it is clear that the environment plays the larger part in the onset of the disease. It is not easy to be sure what this environmental contribution could be, since we would imagine that identical twins would be exposed to the same obvious environmental factors, such as diet, etc. It may be very subtle; for example, an important factor in the development of multiple sclerosis, another autoimmune disease with a clear genetic component, is thought to be a virus infection. The locus labelled *IDDM1* in the HLA region is thought to contribute about 40% of the genetic risk. In European populations, the *IDDM1* genotype DR3/DR4 predisposes towards diabetes, but this can vary in other ethnic groups. The concordance of DR3/DR4 is very high in affected sib-pairs. However, whereas 2% of European populations have the DR3/DR4 haplotype, only 0.1% succumb to diabetes. Moreover, there are some who have IDDM who do not have the DR3/DR4 genotype. Thus the great majority of people who have the risk genotype do not succumb to the disease and some who have the disease do not have the predisposing genotype. It would clearly be wrong to label *IDDM*1 as the 'diabetes gene'. Nevertheless, many independent lines of research confirm that *IDDM*1 contributes towards the risk of diabetes and we can even start to formulate plausible models of its action.

Let us now examine a more controversial topic, the so-called 'gay gene', in the light of the diabetes example. A region of the X chromosome associated with male homosexual orientation (termed a 'gay phenotype') has been identified as follows. Firstly, pedigree studies in a group of families in which a number of males were gay demonstrated clear X-linkage. Allele sharing between gay brothers was then used to identify linkage between loci at Xp28 and homosexual orientation with a high degree of statistical confidence. This observation was then repeated with an independent group of families.

So far this study follows standard methods and had it been a less controversial trait would have been unremarkable. The difficulty comes with the use of the term 'gay gene'. In the first place, no gene has been identified, only a chromosomal region. More importantly, we saw in the case of diabetes the danger in using such a term. Some of the gay brothers did not have a genotype associated with the gay phenotype. We have no idea of the population incidence of whatever genetic element has been identified. Remember that in the case of diabetes only 1 in 20 people with the DR3/DR4 genotype develop diabetes. Remember also that despite the clear evidence of genetic predisposition to diabetes, the environment still plays a large role. We use terms such as 'gay gene' to avoid circumlocutions. It is clumsy to repeat 'a region which is identical by descent in pairs of gay brothers more frequently than would be expected by chance'. However, some forms of contraction may be preferable to others: 'a region associated with a gay phenotype' is better than 'gay gene'.

11.7 Summary

- It is important that advances in human genetics lead to the relief of suffering and the promotion of good health. There are clearly foreseeable ways in which they may have a negative impact on society through genetic discrimination and loss of privacy.

- Genetic tests allow diagnosis of genetic illness. They can be used in a variety of situations to prevent, or allow the early treatment of, diseases that have a genetic component.

- Genetic testing can have negative consequences. It can cause psychological harm or it may lead to actual discrimination in insurance or employment.

- Genetic tests should only be used when a clear benefit can be derived. They must only be used after the person tested has given their informed consent. Informed consent requires counselling.

- Everyone tested has the right to expect confidentiality. This can lead to some conflicts of interest when there are good reasons for other family members to know the result of a test.

- As well as the right to know comes the right not to know. Care must be taken to ensure that unwelcome information is not given inadvertently.

- Insurance companies wish to know about the results of a genetic test to avoid fraud. This results in a disincentive to genetic tests that may be important for someone's health.

- Employers may wish to discriminate on the basis of genetic tests to avoid hiring or promoting those with poor health prospects. Genetic tests may sometimes be justified on the grounds of public safety or to protect those who are especially sensitive to environmental hazards.

- Confidentiality may be compromised by media interest.

- Predisposition to antisocial behaviour raises issues of individual responsibility for criminal actions or substance abuse. Equally, it may lead to particular groups being labelled as genetically inferior. In such arguments, we must never lose sight of the large contribution made by the environment.

- Gene therapy offers the possibility of a cure for genetic diseases. Somatic gene therapy is not thought to raise any ethical problems. Germline gene therapy may have considerable advantages over somatic gene therapy. However, practical difficulties prevent it being used at present and there are powerful ethical arguments against it ever being used; at present germline gene therapy is prohibited.

- Patents in human genetics are required to allow private investment to fund research. A debate over whether genes and DNA sequences can be patented has been settled in favour of allowing such patenting.

- Genes influence our health, personalities and aptitudes, but genetic determinism is misguided. These characteristics come about through the interaction of many genes with each other and with the environment. These interactions are complex and unlikely ever to yield simple predictions about their outcome.

- We must be careful in our choice of shorthand terms to describe complex concepts. Terms such as 'gay gene' convey the wrong meaning to anyone not versed in the complexities of multifactorial inheritance.

Further reading

EISENBERG, R.S. (1997) Structure and function in gene patenting. *Nature Genetics,* **15,** 125–131.

House of Commons Science and Technology Committee Third Report (1995). *Human Genetics: the Science and its Consequences. Volume 1. Report and Minutes of Proceedings.* HMSO, London.

Human Genetics Advisory Commission (1997) The implications of genetic testing for insurance. HGAC Secretariat, Office of Science and Technology, Albany House, 94–98 Petty France, London SW1H 9ST, UK.

Nuffield Council on Bioethics (1993). *Genetic Screening: Ethical Issues.* Nuffield Council on Bioethics, 28 Bedford Square, London WCIB 3EG.

PARKER, L.S. (1995) Ethical concerns in the research and treatment of complex disease. *Trends in Genetics,* **11,** 520–523.

REILLY, P.R. (1997) Fear of genetic discrimination drives legislative interest. Ownership, predisposition major issues. *Human Genome News,* **8** (3). http://www.ornl.gov/hgmis/publicat/hgn/v8n3/01fear.html

SCHMIDTKE, J. (1992) Who owns the human genome? Ethical and legal aspects. *Journal of Pharmacy and Pharmacology,* **44** (Suppl. 1), 205–210.

Promoting safe and effective genetic testing in the United States. Principles and recommendations. Task force on Genetic Testing of the NIH-DOE Working Group on Ethical, Legal, and Social Implications of Human Genome Research.
http://www.med.jhu.edu/tfgtelsi/promoting/

The HUGO Web site.
http://www.ornl.gov/TechResources/Human_Genome/research.html

Glossary

aboriginal population An original or native population so far unaffected by recent mixing with other populations.

acrocentric A chromosome where the centromere is near one end.

additive interaction Where the effect on phenotype of multiple alleles is the sum of their individual effects.

admixture Gene flow between populations as a result of migration.

aetiology The causes of a disease.

affected pedigree member Two genetically related individuals who are both affected by a disease in a pedigree.

affected sib-pair Two sibs affected by a disease.

allele One of several alternative forms of a gene or DNA sequence.

allele frequency The frequency in a population of each allele at a polymorphic locus.

allele-specific oligonucleotide An oligonucleotide that will only anneal to one allele, either mutant or wild type.

Alu The most common repeated sequence in the SINE category.

amniocentesis Sampling of cells from the amniotic fluid surrounding the developing embryo for genetic testing. The procedure involves inserting a large tube through the mother's abdomen. These cells are cultured *in vitro* for about 3 weeks to increase numbers. The procedure cannot be attempted before the 16th week of pregnancy and is now less commonly used than chorionic villous sampling.

amplification refractory mutation system A commercial multiplex PCR system for genetic testing.

amyloid precursor protein Precursor of the amyloid protein found in senile plaques in the brains of Alzheimer's disease patients.

anaemia Blood disorder characterised by deficiency of red blood cells.

anticipation The phenomenon by which the severity or penetrance of a disease apparently increases with each succeeding generation.

antigen A molecule or part of a molecule recognised by an antibody.

anti-onocogene *See* tumour suppressor.

antisense mRNA An RNA molecule or oligonucleotide that is complementary to an mRNA molecule. It can thus form a duplex with the mRNA molecule, interfere with its translation and provoke its destruction by ribonucleases specifc for double-stranded RNA molecules.

APC **(adenomatous polyposis coli)** The gene that dysfunctions in familial adenomatous polyposis.

ApoE *See* apolipoprotein E.

apolipoprotein E Lipoprotein found in blood plasma. Different alleles at the apolipoprotein E gene have been demonstrated to be risk factors for both cardiovascular disease and Alzheimer's disease.

apoptosis Programmed cell death provoked by irreparable genetic damage.

APP *See* amyloid precursor protein.

archaic humans Members of the species *Homo sapiens* that pre-dated anatomically modern humans.

ARMS *See* amplification refractory mutation system.

Ashkenazi Jews A particular population group that originated in eastern France in the early Middle Ages, who subsequently migrated to eastern Europe and later to the USA and Israel. By virtue of culture and religion they have been largely reproductively isolated from surrounding populations.

association Simultaneous occurrence of a disease phenotype and an allele at a frequency that is statistically significant.

ataxia Unsteady gait and uncoordinated movements.

autoantibody An antibody produced by an individual's own immune system that recognises antigens in their own body.

autoantigen The antigen recognised by the body's immune system in an autoimmune disease.

autoimmune Attack by an individual's own immune system on some part of the individual's body.

autosomal dominant A trait showing a characteristic pattern of segregation indicating that the gene in question is located on an autosome and that the mutant allele is dominant to the wild-type allele. Thus the disease is manifested when one copy of the mutant allele is inherited.

autosomal recessive A trait showing a characteristic pattern of segregation indicating that the gene in question is located on an autosome and that the mutant allele is recessive to the wild-type allele. Thus the disease is only manifested when two copies of the allele are inherited.

autosome A chromosome that is not an X or Y chromosome. There are 22 autosomes in the human chromosome complement.

BAC *See* bacterial artificial chromosome.

BAC-end sequencing A strategy to obtain sequence-ready clones from the human genome.

bacterial artificial chromosome A cloning vector that uses the origin of replication of the F plasmid. Can take inserts of 100–300 kb and is thought to be less prone to cloning artefacts, such as chimeric inserts or rearrangements, upon propagation within the host.

balanced polymorphism A polymorphism maintained in the population because the heterozygous state has a greater Darwinian fitness than either homozygous state. Normally, one homozygote is at a severe selective disadvantage, but the allele is maintained in the population at a significant frequency because of the selective advantage of heterozygotes.

balanced translocation When translocated chromosomes are inherited together so there is no change in genomic information content, apart from at the breakpoint.

banding Any process that produces a discrete pattern that can be used to identify individual chromosomes and chromosomal regions. *See* G-bands and Q-bands.

biallelic A locus with two alleles.

bivalent A chromosome with two chromatids.

bp Base pair.

buoyant density ultracentrifugation A process of centrifugation that separates molecules according to density rather than mass. Analysis of DNA molecules involves mixing the DNA with a solution of CsCl whose concentration is chosen to closely match the average density of DNA. The solution is centrifuged at very high speed (40 000–50 000 g) for 48 hours. The centrifugal force generates a gradient in CsCl concentration. The DNA molecules move to an equilibrium position where their density is matched by the density of the CsCl at a particular point in the gradient. The density of DNA is determined by its base composition. Any fraction of the genome whose DNA content differs from the average will form a separate or satellite peak.

CAAT box The sequence GGCCAATCT that can be found in the promoter region of genes and which binds the transcription factors CTF and NF1 to stimulate transcription.

cap Covalent modification of the 5' end of an mRNA molecule during transcription.

capsid The protein coat of a virus particle.

CBAVD *See* congenital bilateral absence of the vas deferens.

cDNA library Gene library composed of cDNA inserts synthesised from mRNA using reverse transcriptase.

cell cycle engine Molecular mechanism that controls the passage of cells through the cell cycle in eukaryotic cells. It consists of one or a family of protein kinase subunits whose activity is regulated by association with different regulatory subunits called cyclins.

centimorgan Unit of genetic map distance corresponding to a recombination fraction of 0.01. Named after Thomas Hunt Morgan.

centromere A constriction visible in metaphase chromosomes where the chromosome is attached to the mitotic or meiotic spindles.

CEPH families Set of reference families for genetic mapping maintained by the Centre d'Étude Polymorphism Humain in Paris.

CF Cystic fibrosis.

CFTR *See* cystic fibrosis transmembrane conductance regulator.

checkpoint A mechanism that prevents a cycle event from occurring if a previous event is not completed.

chorionic villous sampling The chorion is derived from the zygote but is not part of the developing embryo. It is a membrane with projections or villi that surround the embryo. Chorionic villous sampling is carried out by introducing a catheter through the vagina until it touches the chorion. It can be safely carried out in the eighth or ninth week of pregnancy. It thus has an advantage over amniocentesis because it allows an earlier termination of pregnancy should this be indicated by the results of the genetic test.

chromatid A chromosomal strand consisting of a complex of a single DNA molecule and its associated protein and RNA components.

chromatin The complex of DNA, RNA and protein that makes up a chromosome.

chromosome jumping A method used in the course of a chromosome walk that allows tens or hundreds of kilobases to be covered in a single operation.

chromosome walking A method of moving from a linked marker to a gene.

cis *See* phase.

cladistic Method of phylogenetic classification that traces organisms back to a common ancestor using qualitative characters to determine the order of descent.

clone library A collection of clones that ideally contains all possible sequences from a donor genome.

clone map A physical map based on determining overlap between clones.

CMC Chemical mismatch cleavage.

coding sequence A length of DNA or RNA whose sequence determines the sequence of amino acids in a protein.

complement fixation The binding of a group of blood globulin proteins to cause lysis of foreign cells after they have been coated with antibody.

complex disease Diseases whose aetiology consists of a mixture of environmental and genetic factors.

compound heterozygote An individual that lacks a gene function because each copy of the gene is inactivated by different mutations.

concordance The percentage proportion of identical twins who both suffer from a disease or exhibit a phenotypic trait when that disease or trait occurs in one member of the pair.

congenital disorder A disorder present at birth.

congenital bilateral absence of the vas deferens Often one of the symptoms of cystic fibrosis that results in male sterility.

consanguineous The individuals concerned are related and therefore have a proportion of their genes in common.

consensus sequence An idealised sequence that represents the nucleotides most often found at each position when a number of related sequences are compared.

contig A DNA region represented by a group of clones whose relationship one to another is defined by overlap between the sequences of the inserts.

cosmid Cloning vector that takes the form of a plasmid that contains the *cos* site from phage λ. This allows recombinant DNA molecules of ~40 kb to be packaged into λ phage particles *in vitro*. Upon infection into a host cell the recombinant molecule replicates as a plasmid.

CpG island A region of approximately 1 kb in which the CpG dinucleotides are not methylated and whose frequency is higher than that found in the rest of the genome, being equal to that expected from the percentage (G+C) of human DNA. CpG islands often span the promoter region of transcriptionally active genes and sometimes the start of the coding sequence.

CRE Cyclic AMP response element.

cystic fibrosis transmembrane conductance regulator The protein affected by mutation in cystic fibrosis patients. Also used to describe the encoding gene.

cytogenetic A characteristic of the karyotype that is revealed by examination of metaphase chromosomes.

cytokine Hormone-like factor that regulates the activity of the immune system.

Darwinian fitness The number of an organism's offspring reaching reproductive maturity.

deletion mutations The loss of one or more nucleotides.

diabetes mellitus Syndrome caused by failure of cells to take up glucose.

diastrophic dysplasia Hereditary disease characterised by bilateral clubbed feet, malformations of the outer ear caused by calcification of the cartilage, premature calcification of the costal cartilage, sometimes cleft palate, and characteristic malformation of the thumb dubbed 'hitch-hiker's thumb'.

dinucleotide Two successive nucleotides on the same DNA strand, written with the 5' nucleotide first.

distal Away from the centromere compared with the chromosomal point of reference.

dizygotic twins Twins derived from different fertilised eggs. Otherwise known as non-identical twins.

DMD Duchenne muscular dystrophy.

DMPK *See* dystrophia myotonica protein kinase.

DNA chip A high-density miniaturised array of oligonucleotides attached to silica or glass substratum. They are being developed to scan genes for the presence of mutations. It may be possible to use them for automatic sequencing.

DNA fingerprinting Any method that identifies unique features of a clone which can be used to determine overlap between other clones in a library.

DNA methylation The addition of a methyl group to cytosine to form 5-methylcytosine in CpG dinucleotides.

DNA profile Features of a DNA sample that can be used to identify the individual from which it originated. Differs from genetic fingerprinting in that it does not purport to uniquely identify the individual from all other humans.

domain Part of a protein where a continuous length of the polypeptide chain folds to form a discrete globular structure that may have a defined function.

dominant negative mutation A mutation that prevents another wild type protein in the same cell from functioning. Commonly acts by producing an altered polypeptide that prevents the assembly of a multimeric protein.

downstream A region of the DNA molecule that lies 3' to the point of reference.

dynamic mutation A trinucleotide repeat expansion mutation that changes in size during meiosis or, in some cases, mitosis.

dystrophia myotonica *See* myotonic dystrophy.

dystrophia myotonica protein kinase Implicated in myotonic dystrophy, as the trinucleotide repeat expansion responsible for the disease occurs in the 3' untranslated region of this gene.

dystrophin The protein affected by mutation in Duchenne muscular dystrophy. Also used to describe the encoding gene.

EMC Enzyme mismatch cleavage.

endocytosis A process by which eukarytotic cells take in material from the outside. The cell membrane invaginates to form intracellular vesicles, which then fuse to form the endosome.

endosome *See* endocytosis.

enhancer A *cis*-acting DNA sequence that increases the expression of a gene upon binding a transcription factor. The action of an enhancer is not critically dependent on its position or orientation and in many cases exerts its influence at a considerable distance.

episome A DNA molecule that can stably replicate. The significant point is that it does not have to be integrated into a chromosomal location to be propagated in daughter cells.

epistasis Originally defined as the interaction between two genes so that the expression of one is controlled by the other. Now more loosely applied to situations where the effect on phenotype of alleles at two genes is different from that which would be expected by combining the individual effect of each allele.

erythropoiesis Production of new red blood cells.

EST *See* expressed sequence tag.

euchromatin The main fraction of chromosomal DNA that is not heterochromatin. It is uncoiled during interphase and contains transcriptionally active regions.

eugenics 'Improvement' of the human genetic stock by selective breeding.

exon The part of a gene that is transcribed and remains in an mRNA molecule after splicing. This mainly consists of protein-coding regions but also includes 5' and 3' untranslated regions of the mRNA molecule.

exon trapping Special technique used to search for exons in genomic clones.

expressed sequence map A map that plots the position of DNA sequences expressed in mRNA. Based on EST markers.

expressed sequence tag A sequence tagged site derived from a cDNA clone. Used to map the positions of expressed sequences.

expressivity The difference in the severity of a disorder in individuals who have inherited the same disease alleles.

ex vivo **gene therapy** Where cells are removed from the body, manipulated *in vitro* and then reintroduced to the patient.

familial Where a disease is transmitted in families as opposed to its sporadic occurrence.

familial adenomatous polyposis Cancer of the colon and rectal areas. Inherited as an autosomal dominant trait due to mutations in the APC gene.

FAP *See* familial adenomatous polyposis.

Feulgen *See* heterochromatin.

FISH *See* Fluorescence *in situ* hybridisation.

fluorescence *in situ* hybridisation *In situ* hybridisation in which the probe is labelled with a fluorophore. Upon examination with a fluorescence microscope, the chromosomal region to which the probe is binding can be visualised.

fluorophore A chemical moiety that fluoresces when stimulated by light of a particular wavelength, usually in the ultraviolet range.

founder effect The effect that results in the genetic structure of the population derived from the founder group being different from the ancestral population.

founder mutation A disease-causing mutation on chromosomes in different individuals that have descended from an ancestral chromosome on which the original mutation occurred.

founder population A small group of individuals that splits off from the main population and settles in previously uninhabited territory.

fragile site Non-staining gap in a chromosome visible in metaphase spreads of cells cultured under conditions such as folate deprivation or chemical inhibition of DNA synthesis.

frameshift mutation Deletion or insertion of bases that alter the reading frame.

G_1 Gap 1. The period between nuclear division and DNA synthesis (S phase) in the cell cycle.

G_2 Gap 2. The period between DNA synthesis (S phase) and nuclear division.

gametogenesis The process by which gametes are produced.

ganglioside A group of complex glycolipids found chiefly in nerve membranes, containing sphingosine, fatty acids and an oligosaccharide chain containing at least one acid sugar, such as N-acetylglucosamine, N-acetyl neuramic acid and N-acetylgalactosamine.

gap junction A small area where two adjacent cells are joined by a continuous aqueous channel through their plasma membranes.

G-bands Bands produced in metaphase chromosomes using Giemsa.

GC box The sequence GGGCGG found in the promoter region of genes, which binds the transcription factor SP1 to stimulate transcription.

gene A DNA sequence that contributes to the phenotype of an organism in a way that depends on its sequence. Normally this refers to a protein-coding sequence, together with any sequences that are required for expression. Some genes encode RNA that is functional directly and is not translated.

gene families Similar but not identical genes that have arisen by a process of gene duplication and divergence. Gene families can be dispersed through the genome or exist as gene clusters at one location.

gene library *See* clone library.

general transcription factor A protein required for the formation of the transcription complex. Normally designated in the form TFIIX, where the roman numeral indicates the RNA polymerase (in this case RNA polymerase II) and X denotes the particular protein.

gene therapy Treatment of a disease by genetic modification of the patient's cells.

Genethon Laboratory in Paris funded by the French muscular dystrophy association. Constructed the standard genetic map of the human genome.

genetic drift The change in allele frequencies from one generation to another due to sampling effects.

genetic heterogeneity Where apparently clinically similar disorders are caused by mutations in different genes (non-allelic), or where mutations in the same gene result in clinically diverse conditions (allelic).

genetic map A genomic map based on the order of genetic mapping markers and the genetic distance between them measured by the recombination frequency.

genome The sum total of the genetic material of an organism.

genome scan A systematic survey to discover if a phenotypic trait or genetic disease is linked to a genetic mapping marker.

genomic imprinting A mechanism whereby cells preferentially express an autosomal gene inherited from one parent.

germline The cells that will give rise to gametes. Mutations in germline cells will be inherited by the next generation.

germline gene therapy Where the genetic modification may affect the germline and be passed on to succeeding generations.

Giemsa A dye that binds DNA.

glycolipid A lipid containing a carbohydrate chain.

Golgi apparatus A membranous cellular compartment that forms part of the secretory pathway.

growth factor A specific factor that must be present in the environment before a cell can proliferate. Usually a protein or peptide that interacts with a growth factor receptor at the cell surface.

haemoglobinopathies A group of genetic diseases that affect the production or function of haemoglobin.

haplotype A set of closely linked alleles that tend to be inherited together, i.e. not separated by recombination at meiosis.

Hardy–Weinberg distribution The mathematical description of the distribution of genotypes at a particular locus in a population.

HBC *See* hereditary breast cancer.

HD Huntington's disease.

hemizygous The presence of only one copy of a gene. Usually applied to genes on the X chromosome in males.

hereditary breast cancer A subgroup of breast cancer cases where the cancer is caused by the inheritance of an autosomal dominant allele of genes such as *BRCA1* and *BRCA2*.

hereditary non-polyposis colorectal Cancer of the colon and rectal areas. Inherited as an autosomal dominant trait due to mutations in the APC gene.

heritability The proportion of total population variability that can be accounted for by genetic variability.

heterochromatin Chromosomal regions that are transcriptionally inactive and stain more densely with Feulgen, a dye that binds DNA.

heteroplasmic Where both mutant and wild-type mitochondrial genomes are found in the same individual or cell.

heterozygous The presence of two different alleles at a specified locus.

heterozygous advantage When an allele that is deleterious in the homozygous state is advantageous in the heterozygous state.

highly repetitive DNA Simple sequences repeated up to 10^6 per genome.

histones Highly conserved basic proteins that bind DNA and organise the formation of nucleosomes.

HLA *See* human leucocyte antigens.

HNPCC *See* hereditary non-polyposis colorectal cancer.

homologous genes or DNA sequences Genes whose sequence similarity in two different species suggests a common evolutionary origin.

homoplasmic The state where all mitochondrial DNA in an individual or cell is identical.

homozygous The presence of two identical alleles at a specified locus.

housekeeping gene A gene that encodes a protein performing a basic function common to most cells.

HRE Heat-shock response element.

human leucocyte antigens Antigens found on the surface of cells that signal whether the cell is self or non-self.

huntingtin The protein affected by mutation in Huntington's disease. Also used to describe the encoding gene.

hybridisation The formation of a hybrid DNA molecule (or RNA/DNA duplex) by the base pairing of complementary strands that originated from different sources.

hydrops fetalis A condition lethal at or before birth caused by a complete absence of α-globin.

IBD *See* identical by descent.

IBS *See* identical by state .

IDDM *See* Insulin-dependent diabetes mellitus.

identical by descent A situation where two individuals in a kindred have inherited the same alleles at a locus from a common ancestor.

identical by state A situation where two individuals in a kindred have the same alleles at a locus but it cannot be demonstrated that they are identical by descent.

immortal The property of a cancer cell line that enables it to be propagated indefinitely in culture.

informative meiosis A meiosis where it is possible to distinguish between parental and recombinant chromosomes in the progeny.

in situ **hybridisation** Hybridising a nucleic acid probe to chromosomes spread on a slide. Usually a metaphase spread is used but recently also carried out with interphase spreads to increase resolution.

insulin-dependent diabetes mellitus Also known as type 1 diabetes. Form of diabetes mellitus that responds to insulin. Usually shows juvenile onset and is caused by the autoimmune destruction of the islets of Langerhans in the pancreas.

intermediate repeated DNA Sequences repeated between 100 and 10^5 times per genome.

interphase The period of the cell cycle between nuclear divisions.

intron The portion of a gene that is transcribed but removed during splicing.

in vivo **gene therapy** Where the transgene is introduced directly into the target cells of the patient's body.

karyotype The number, size and shape of chromosomes in a somatic cell.

kb Kilobase. A thousand base pairs.

ketone bodies Group of chemical compounds, such as β-hydroxybutyrate, acetoacetone and acetone, that are metabolites of acetyl-CoA and which accumulate in diabetes.

ketonemia Accumulation of ketone bodies in the bloodstream in diabetes.

ketonuria Accumulation of ketone bodies in the urine in diabetes.

ketosis Accumulation of ketone bodies in diabetes.

kindred A group of people that are related genetically or by marriage.

kinetechore A protein structure that assembles at the centromere to which the mitotic or meiotic spindle attaches.

knockout mice Mice where specific genes have been deleted.

kya Thousand years ago.

λ 1. Relative risk. 2. Bacteriophage λ. A commonly used vector for gene cloning.

L1 A repeated sequence in the LINE category.

lariat A splicing intermediate where the intron resembles a stem–loop structure or lasso.

liability The combined total of genetic and environmental effects that predispose an individual to a multifactorial disease.

LINEs *See* long interspersed elements.

linkage The propensity of two genetic markers to be co-inherited through meiosis. It indicates that the markers are located close together on the same chromosome.

linkage disequilibrium The tendency of particular alleles at one locus to be associated with particular alleles at a second closely linked locus.

lipoplex Complex formed between cationic lipids and DNA used for gene therapy.

liposome Spheres consisting of lipid molecules surrounding an aqueous interior. Used to encapsulate DNA for gene therapy.

locus A unique chromosomal region that corresponds to a gene or some other DNA sequence.

LOD score The logarithm of the ratio of odds that two loci are linked with a specified recombination fraction θ to the odds that they are unlinked. An LOD score of 3 or more is required for linkage to be significant.

LOH *See* loss of heterozygosity.

long interspersed elements A category of intermediate repeated sequence.

long range restriction map *See* rare cutting enzyme.

long terminal repeat Repeated sequences at each end of a retrovirus.

loss of heterozygosity When a region of a chromosome becomes homozygous for alleles that were heterozygous.

LTR *See* long terminal repeat.

Lyonisation *See* X-chromosome inactivation.

lysosome A membrane-bound organelle that contains digestive enzymes that break down material taken in by phagocytosis or recycle cellular components.

major disorder A disorder that is both relatively common and severe in its consequences.

maturity-onset diabetes of the young A form of non-insulin-dependent diabetes that affects teenagers and young adults.

maximum likelihood score The most likely of a series of alternatives judged by which gives the highest LOD score. For example, the maximum likelihood score for different values of recombination fraction, θ, indicates the most likely map distance between two loci.

Mb Megabase. One million base pairs.

MD Myotonic dystrophy.

meconium ileus Obstruction of the bowel in the newborn. Often one of the symptoms of cystic fibrosis.

meiosis A process consisting of two successive nuclear divisions that results in reduction of chromosome number from the diploid chromosome complement to the haploid complement. During the first meiotic prophase recombination takes place between homologous chromosomes.

Mendelian inheritance The pattern of segregation of a disease in a family or kindred that conforms to Mendel's laws of inheritance. Normally taken to indicate that the disease is monogenic.

mesolithic A period of the stone age between the palaeolithic and neolithic periods.

metacentric A chromosome that has a centrally located centromere, such that both arms are of equal size.

microsatellites Tandem repeats of a short sequence 2–4 bp in length found at many different locations in the genome.

minisatellites A class of highly repetitive sequences that consist of sequences between 10 and 100 bp long repeated in tandem arrays which vary in size from 0.5 to 40 kb. They tend to occur near telomeres, although they have been found elsewhere. Also known as variable number tandem repeats (VNTRs).

minisatellite variant repeat PCR A form of PCR that analyses variation within the sequences of minisatellite repeat units.

missense mutation Nucleotide substitution that results in an altered amino acid sequence in the encoded protein.

mitogen A substance that stimulates cell division.

mitosis The process of nuclear division in which the daughter cells contain the same number of chromosomes as the mother cell.

mitotic crossing-over Recombination that occurs in mitosis of somatic cells. In daughter cells results in homozygosity of the region of the chromosome distal to the cross-over point.

MLS *See* maximum likelihood score.

mobile genetic element A sequence of DNA that can move from one chromosomal location to another by transposition. Characteristically they are flanked by short direct repeats.

monogenic *See* single-gene defect.

monosomic A somatic cell with one copy of a chromosome.

monozygotic twins Twins derived from the same fertilised egg; otherwise known as identical twins.

mortal The propensity of human cells to die after about 50 generations in tissue culture.

mosaicism When not all cells in the body are genetically identical. This may come about through a mutation in early development and may result in either the germline or somatic cells being affected.

multifactorial disease A disease caused by the interaction of the environment and polygenes or oligogenes.

multiplex PCR Where several PCR reactions are carried out together using more than one pair of primers.

multiplicative interaction Where the effect on phenotype of multiple alleles is the product of their individual effects. Multiplicative interaction is a special case of epistasis.

multipoint mapping Testing for linkage between genetic markers and a disease by using more than one marker at a time.

multiregional evolution hypothesis Theory of human evolution which states that *Homo sapiens* evolved independently from *Homo erectus* on multiple occasions in different parts of the world.

MVR-PCR *See* minisatellite variant repeat PCR.

mya Million years ago.

myoblast Mononucleate cell that is normally quiescent but can divide and fuse with muscle fibres to repair damage.

myotonic dystrophy Neuromuscular disorder caused by a trinucleotide repeat expansion.

neolithic Sometimes called the new stone age. A period of human history which started about 10 000 years ago marked by the appearance of agriculture, settled communities and polished stone tools.

neural crests A ridge of ectoderm that forms above the neural tube during early embryogenesis. Subsequently the cells migrate and develop into the dorsal root ganglia of the sensory nervous system, the adrenal medulla and some skeletal elements of the face.

neutral allele One of the alleles of a neutral polymorphism.

neutral molecular polymorphism A polymorphism for a DNA sequence that has no effect on the genotype.

neutral polymorphism A polymorphism in which the different alleles do not have a discernible effect on the fitness of the organism.

neutral theory of evolution Theory which postulates that the main mechanism generating nucleotide substitution is the fixation of neutral alleles through the effect of genetic drift.

NIDDM *See* non-insulin-dependent diabetes mellitus.

NOD *See* non-obese diabetic mouse.

non-disjunction The failure of homologous chromosomes to segregate at meiosis, resulting in one daughter cell with two copies and one daughter cell with no copies of the chromosome in question.

non-insulin-dependent diabetes mellitus Form of diabetes that does not respond to insulin. Typically shows onset in middle age.

non-obese diabetic mouse A strain of mouse that spontaneously develops a disease that closely resembles IDDM in humans.

non-parametric A method of mapping disease genes that does not require the formulation of models with specified parameters.

nonsense codon A codon that does not encode an amino acid and causes translation to terminate.

nonsense mutation Nucleotide substitution that results in a nonsense codon, which results in a truncated protein. Also result from frameshift mutations, as the incorrect reading frame will contain nonsense codons.

non-shared environment The combined non-genetic components that result in identical twins reared together having different phenotypes.

Northern blotting A procedure similar to Southern blotting but where RNA not DNA is fractionated and hybridised to the probe. Allows the size and expression of mRNA molecules to be studied.

nucleosome Basic unit of DNA packing in which DNA winds around a tetramer of histones H2A, H2B, H3 and H4 to form a beads-on-a-string appearance in the electron microscope.

Oct box The sequence ATTTGCAT that can be found in the promoter region of genes, which binds the transcription factors Oct-1 or Oct-2 to stimulate transcription.

oligogene A gene which makes a large contribution to the aetiology of a disorder.

oligogenic disorder A disorder caused by the action of a few genes.

oligonucleotide A polymer of 5–100 nucleotides.

oncogene A gene the mutation of which is implicated in the aetiology of cancer.

oogonia The diploid precursors of female gametes.

open reading frame A significantly long sequence of codons in one of the six possible reading frames in a DNA sequence that does not contain a nonsense codon.

ordered clone library A clone library with overlapping inserts where overlaps between all the clones have been determined.

ORF *See* open reading frame.

Out-of-Africa hypothesis Theory of human evolution which states that modern humans are descended from a small group that emerged from Africa 100 000 years ago.

Oxidative phosporylation The production of ATP at the inner membrane of the mitochondria. Electrons are passed along the respiratory chain in the mitochondria, eventually interacting with oxygen and protons to form water. The energy released in the process is used to pump protons across the inner mitochondrial membrane. As the protons re-enter the mitochondrial matrix the energy of the proton gradient drives the synthesis of ATP.

P1-derived artificial chromosome A cloning vector based on generalised transduction using phage P1. Can accept insets of approximately 100 kb and is liable to fewer cloning artefacts than YAC vectors.

PAC *See* P1-derived artificial chromosome.

palaeolithic Sometimes called the old stone age. Phase of human history prior to the advent of agriculture and settled communities. Marked by a hunter-gatherer lifestyle and chipped stone tools.

parametric A model of inheritance that can be tested by LOD score analysis that requires the specification of parameters.

parasitic DNA *See* selfish DNA.

p arm The shorter arm of a chromosome in a metaphase spread.

pathophysiology The biochemical or cellular changes that cause the symptoms of a disease.

PC *See* principle component analysis.

PCR *See* polymerase chain reaction.

pedigree A representation of the ancestral relationship between individuals related genetically or by marriage.

penetrance The frequency with which a particular genotype manifests itself in the phenotype.

PFGE See pulsed field gel electrophoresis.

phase Specifies whether particular alleles at adjacent loci are on the same (*cis*) or different (*trans*) chromosomes.

phenocopy The occurrence of a disorder caused by an environmental factor with the same symptoms as an inherited disorder.

phylogenetic tree An attempt to describe the history of populations or species since they diverged from a common ancestor. The order in which the branches split describes the order in which populations diverged, and the length of the branches describes the time since the divergence.

physical map A genomic map based on the actual location of DNA sequences.

PIC *See* polymorphism information content.

polygene One of several genes each of which makes a small contribution to the aetiology of a disorder.

polygenic disorder A disorder caused by the action of several genes.

polymerase chain reaction. A technique for amplifying DNA *in vitro* using a thermostable DNA polymerase and primers that anneal at sites which flank the region to be amplified.

polymorphism information content Measure of the informativeness of a genetic marker.

polymorphic A locus that is heterozygous at a significant frequency in a population.

population bottleneck When a previously large population contracts to a small number and then expands to its previous number. As a result the gene pool may be less diverse and gene frequencies may be altered because of the effects of sampling and genetic drift.

positional candidate cloning Cloning a disease gene by identifying a biologically relevant gene located near the map position of the disease gene.

positional cloning The cloning of a gene using no other information apart from its chromosomal location.

premutation A mutation that has no phenotypic effects but which increases the odds of a trinucleotide repeat expansion in a subsequent generation.

primary RNA transcript The RNA molecule before splicing, which contains both introns and exons.

primers Oligonucleotides that anneal to template DNA to prime synthesis mediated by DNA polymerase.

principal component analysis Mathematical procedure to represent multi-variate data by producing a synthetic map.

private mutation A mutation unique to a particular family or individual.

proband An individual suffering from a genetic disease who is the first member of a pedigree to come to medical attention.

processed pseudogene. A pseudogene that lacks introns and is therefore thought to have arisen by retrotransposition.

promoter Part of the gene upstream of the transcribed region that contains the elements necessary for a basal level of transcription.

protein kinase An enzyme that phosphorylates another protein and so regulates its activity.

proteome The complete set of proteins that can be encoded by a genome.

proximal Towards the centromere compared with the chromosomal point of reference.

pseudogene A DNA sequence that closely resembles a functional gene but which has been rendered non-functional by an inactivating mutation.

pulsed field gel electrophoresis A special adaptation of agarose gel electrophoresis, where the direction of the electric field is periodically changed. Used to construct long-range restriction maps. *See also* rare cutting enzyme.

q arm The longer arm of a chromosome in a metaphase spread.

Q-bands Bands produced in metaphase chromosomes using the fluorophore quinacrine.

QTL *See* quantitative trait locus.

quantitative trait locus A locus contributing towards a quantitative phenotype.

quinacrine A dye that binds DNA in chromosomes and fluoresces in ultraviolet light.

radiation hybrid mapping Method of mapping based on the frequency with which STS markers are co-retained on the same genomic fragment after fragmentation by X-rays.

rare cutting enzyme A restriction enzyme that has target sites in human DNA that are more widely spaced than normal. Used in conjunction with pulsed field gel electrophoresis to construct long-range restriction maps, i.e. maps that extend over hundreds of kilobases.

ras A 21-kDa protein with GTPase activity that acts as a molecular switch in signal transduction pathways involving external mitogenic stimulation. The *ras* gene is a protooncogene because mutations that lock it into its active state cause cellular transformation. Mutations to this gene have been implicated in a wide variety of cancers.

reassociation The re-formation of a duplex DNA molecule after the original molecule has been denatured by heat or other physical treatment.

recombinant protein A protein produced by the expression of a gene in a heterologous host cell.

recombination fraction The proportion of chromosomes that are recombinant, i.e. of a different constitution from parent chromosomes.

recurrence risk *See* relative risk.

relative risk (λ) The ratio of the frequency of a multifactorial disease in the relative of an affected person compared with its rate in the general population.

replacement substitution A nucleotide substitution that results in an amino acid substitution in the encoded protein.

response element A region upstream of a gene that activates its transcription in response to a general environmental influence such as heat, serum or cyclic AMP.

restriction fragment length polymorphism A genetic marker based on the presence or absence of a target for a restriction enzyme due to a polymorphism at a single base pair.

retinoblast Retinal epithelial cells that give rise to neuroblasts and neuroglial precursor cells.

retinoblastoma A tumour of the eye originating from the retinal cells. May be sporadic or inherited due to mutation of the *RB* gene.

retrotransposition A process of transposition in which RNA serves as a template for reverse transcriptase. The resulting cDNA is then inserted into the genome at a new location.

retrovirus A virus that has an RNA genome. The mature virion contains reverse transcriptase encoded by the RNA genome. This copies the RNA into cDNA, which is then inserted into the host genome.

reverse transcriptase (RNA-dependent DNA polymerase) An enzyme that copies RNA into DNA; the product is known as cDNA.

RFLP *See* restriction fragment length polymorphism.

RH mapping *See* radiation hybrid mapping.

RT-PCR A form of PCR that amplifies RNA by first converting it to cDNA using reverse transcriptase.

sarcolemma The membranous sheath that surrounds muscle fibres.

satellite DNA *See* highly repetitive DNA.

satellite peak *See* buoyant density ultracentrifugation.

segregation The transmission of alleles through a pedigree in a way that follows Mendel's laws of inheritance.

selection The process that leads to increase in the frequency of a favourable allele and a decrease in the frequency of a deleterious allele.

selfish DNA DNA that does not contribute to the phenotype of the organism, but evolves to increase its copy number in the genome by transposition.

sequence tagged site A site that is identified from its sequence using a PCR reaction. Can be used as a marker in physical, genetic and genomic maps.

serotype The subdivision of bacteria or viruses based on their antigenic properties.

short interspersed elements A category of intermediate repeated sequence.

short tandem repeat An array of short sequences each normally 2–4 bp nucleotides in length. *See also* minisatellites.

sibs Individuals who have the same parents, e.g. brother and sister.

silencer A region upstream of a gene that turns off transcription in response to a particular signal

silent substitution A nucleotide substitution that does not result in an amino acid substitution in the encoded protein because of the redundancy in the genetic code.

SINEs *See* short interspersed elements.

single-gene defect A disorder or disease that is caused by a mutation in a single gene.

single sequence DNA A sequence of DNA that is not repeated in the genome.

somatic Cells in the body that are not part of the germline.

somatic gene therapy Where the attempt to repair the genetic defect is restricted to the somatic cells of a person affected by the disease.

Southern blotting A form of hybridisation where DNA from the target is digested with restriction enzymes and fractionated by agarose gel electrophoresis. The DNA is transferred by capillary action to a nitrocellulose or nylon membrane. The membrane is then hybridised to a radioactively labelled probe and the molecular weight of the hybridising fragments revealed by autoradiography.

spacer region Tracts of DNA that apparently have no function and which separate one gene from the next.

S phase The period in the cell cycle during which nuclear DNA is replicated.

splice acceptor site The junction between the dinucleotide AG at the end of an intron and the start of the next exon.

splice donor site The junction between the end of an exon and the dinucleotide GT at the start of the next intron.

splicing The process by which introns from the primary transcript are removed in the nucleus. The product of splicing of an RNA polymerase II transcript is the mRNA molecule which only contains introns.

SRE Serum response element.

STR *See* short tandem repeat.

stringency During Southern hybridisation duplex formation can still occur even if there is a mismatch between the sequence of the probe and the target sequence. The amount of mismatch that can be tolerated is dependent on the physiochemical conditions during the hybridisation process. For probes longer than about 200 bp, low salt and high temperature (65°C) are high-stringency conditions that only allow duplex formation if the match between the probe and target is perfect or near perfect. Higher salt and lower temperature allow a greater degree of mismatch to be tolerated and are said to be low-stringency conditions. If the probe is an oligonucleotide, the exact conditions for high-stringency hybridisation are calculated from its length and base content.

STS *See* sequence tagged site.

STS content map A map based on the order of sequenced tagged sites in an ordered clone library.

submetacentric A chromosome with the centromere nearer one end compared to the other. Often appears J-shaped in metaphase spreads.

synteny Conservation of gene order on the chromosomes of related species.

synthetic map *See* principal component analysis.

tandem repeat array An identical or near-identical sequence repeated so that each copy lies immediately adjacent to the next and arranged so that the end of one unit abuts the start of the next.

TATA-binding protein The protein that recognises the TATA box to initiate formation of the transcription complex.

TBP *See* TATA-binding protein.

TDT *See* transmission disequilibrium test.

telocentric A chromosome where the centromere is at one end.

telomerase The enzyme that adds the telomere. It contains an RNA molecule that serves as the template for DNA synthesis.

telomere A special structure at the ends of a chromosome that in humans consists of a tandem repeat of the sequence TTAGGG and ends in a 3' extension.

thalassaemia Anaemia caused by an imbalance of globin proteins.

tiling path The set of clones that represents the sequence of a region or entire chromosome with the minimum overlap.

TNF Tumour necrosis factor.

topology The geometrical arrangement of sequences in clones.

trans *See* phase.

transcription factor A protein that binds upstream of a gene to facilitate or stimulate its transcription.

transgene A DNA molecule introduced into a cell to alter its genetic constitution.

transgenic An organism that has a genome modified by the introduction of a transgene.

translocation An exchange of segments between non-homologous chromosomes.

transmission disequilibrium test Ascertains whether a parent who is heterozygous for an associated and a non-associated allele transmits the associated allele more often to affected offspring.

transplacement An organism where genes have been replaced with genes carrying specific mutations.

transposition The process by which a mobile genetic element copies itself and inserts the copy in a new location.

TRE *See* trinucleotide repeat expansion.

trinucleotide repeat expansion A mutation caused by the increase in the number of copies of a repeated trinucleotide.

trisomic A somatic cell with three copies of a chromosome.

tropism The propensity of certain viruses to infect particular cell types.

tumour suppressor Gene that negatively regulates cell division. Mutations to the gene are recessive at a cellular level but show a dominant pattern of inheritance.

type 1 diabetes *See* insulin-dependent diabetes mellitus.

type 2 diabetes *See* non-insulin-dependent diabetes mellitus.

unbalanced translocation The non-reciprocal duplication of a chromosome segment.

uniform resource locator An identifier of a Web site. Usually starts with http:// indicating that it uses hypertext transmission protocol, the standard language used to communicate within the Web.

univalent A chromosome with one chromatid.

untranslated region The region of the mRNA molecule that is not translated. The 5' untranslated region is the part of the mRNA molecule upstream of the AUG start codon. The 3' untranslated region is the part of the mRNA molecule downstream of the stop codon.

upstream A region of the DNA molecule that lies 5' to the point of reference.

URL *See* uniform resource locator.

UTR *See* untranslated region.

vehicle The physical means by which a transgene is introduced into a target cell in gene therapy.

virion A complete virus particle.

VNTR *See* minisatellites.

X chromosome One of the pair of sex chromosomes. XX individuals are female, XY are male.

X-chromosome inactivation The clonal inactivation of one of the X chromosomes in females during development.

X-linked A trait showing a characteristic pattern of segregation indicating that the gene in question is located on the X chromosome.

YAC *See* yeast artificial chromosome.

Y chromosome The male sex chromosome. *See* X chromosome.

Yeast artificial chromosome A cloning vector using a yeast host cell that can accept very large inserts of DNA (~1 Mb).

Zoo blotting Southern blotting where human DNA is used to hybridise DNA from a range of other mammals.

Index